ACE YOUR MIDTERMS & FINALS

INTRODUCTION TO PHYSICS

ACE YOUR MIDTERMS & FINALS

INTRODUCTION TO PHYSICS

ALAN AXELROD, Ph.D.

McGraw-Hill

New York   San Francisco   Washington, D.C.   Auckland   Bogotá
Caracas   Lisbon   London   Madrid   Mexico City   Milan
Montreal   New Delhi   San Juan   Singapore
Sydney   Tokyo   Toronto

Library of Congress catalog card number: 99-70655

McGraw-Hill

*A Division of The **McGraw·Hill** Companies*

1 2 3 4 5 6 7 8 9 0 DOC/DOC 9 0 9 8 7 6 5 4 3 2 1 0 9

ISBN 0-07-007010-5

The sponsoring editor for this book was Barbara Gilson, the editing supervisor was Paul R. Sobel, and the production supervisor was Modestine Cameron. It was set in Minion by Carol Norton for The Ian Samuel Group, Inc.

Printed and bound by R. R. Donnelley & Sons Company.

McGraw-Hill books are available at special quantity discounts to use as premiums and sales promotions, or for use in corporate training sessions. For more information, please write to the Director of Special Sales, McGraw-Hill, 11 West 19th Street, New York, NY 10011. Or contact your local bookstore.

 This book is printed on recycled, acid-free paper containing a minimum of 50% recycled, de-inked fiber.

CONTENTS

PART THREE
MIDTERMS AND FINALS

PART FOUR
FOR YOUR REFERENCE

HOW TO USE THIS BOOK

YOU KNOW THE DRILL. FIRST DAY IN A SURVEY, INTRODUCTORY, OR "CORE COURSE": The professor talks about grading and, saying something about the value of the course in a program of "liberal education," declares that what he or she wants from the students is original thought and creativity and, above all, does not "teach for" the midterm and final.

Nevertheless, the course certainly *includes* one or two midterms and a final, which account for a very large part of the course grade. Maybe the professor can disclaim with a straight face *teaching* for these exams, but few students would deny *learning* for them.

True, you know that the purpose of an introductory course is to gain a useful familiarity with a certain field and *not* just to prepare for and do well on a couple or three exams. Yet, the exams *are* a big part of the course, and, whatever you learn or fail to learn in the course, your performance as a whole is judged in large measure by your performance on these exams.

So the cold truth is this: More than anything else, curriculum core courses *are* focused on the midterm and final exams.

Now, traditional study guides are outlines that attempt a bird's-eye view of a given course. But *Ace Your Midterms and Finals: Introduction to Physics* breaks with this tradition by viewing course content through the magnifying lens of ultimate accountability: the course exams. The heart and soul of this book consists of 11 midterms and 11 finals prepared by *real* instructors, teaching assistants, and professors for *real* students in *real* schools.

Where did we get these exams? Straight from the professors and instructors themselves.

◆ All exams are real and have been used in real courses.

◆ All exams include critical how-to tips and advice from the creators and graders of the exams.

◆ All exams include actual answers.

Let's talk about those answers for a minute. In most cases, the answers are actual student responses to the exam. In all cases, the answers included are A-level responses, and you'll find full commentary that points out why the response is successful.

This book also contains more than the exams themselves.

◆ "Part One: Preparing Yourself" provides you with how-to guidance on what physics professors look for, how to think like a physicist, how to study more effectively, and how to gain the performance edge when you take an exam.

◆ "Part Two: Study Guide" presents a quick-and-easy overview of the content of typical surveys of physics. It clues you in on what to expect in these courses.

◆ "Part Three: Midterms and Finals" is the exams themselves, grouped by college or university.

◆ "Part Four: For Your Reference" provides you with a handy glossary of key terms in physics and a brief list of recommended reading.

What This Book *Is Not*

Ace Your Midterms and Finals: Introduction to Physics offers a lot of help to see you through to success in this important course. But (as you'll discover when you read Part One) the book *cannot* take the place of:

◆ Doing the assigned reading

◆ Working assigned homework problems

◆ Working extra practice problems

◆ Keeping up with your work and study

◆ Attending class

◆ Taking good lecture notes

◆ Thinking about and discussing the topics and issues raised in class and in your books

Ace Your Midterms and Finals: Introduction to Physics is not a substitute for the course itself!

What This Book *Is*

Look, it's both cynical and silly to invest your time, brainpower, and money in a college course just so that you can ace a couple of exams. If you get A's on the midterm and final, but come away from the course having learned nothing, you've failed.

We don't want you to be cynical or silly. The purpose of introductory, survey, or "core" courses is to give you a panoramic view of the knowledge landscape of a particular field. The primary goal of the college experience is to acquire more than tunnel intelligence. It is to enable you to approach whatever field, profession, or work you decide to specialize in from the richest, broadest perspective possible. College is education, not just vocational training.

We don't want you to "study for the exam." The idea is to study for "the rest of your life." You are buying knowledge with your time, your brains, and your money. It's an expensive and valuable commodity. Don't leave it behind you in the classroom at the end of the semester. Take it with you.

Even the starriest-eyed idealist can't deny that midterms and finals are a big part of intro courses and that even if your ambitions lie well beyond these exams (which they should!), performing well on them is necessary to realize those loftier ambitions.

Don't, however, think of midterms and finals as hurdles—obstacles—you must clear to realize your ambitions and attain your goals. The exams are there. They're real. They're facts of college life. You might as well make the most of them.

Use the exams to help you focus your study more effectively. Most people make the mistake of confusing *goals* with *objectives*. Goals are the big targets, the ultimate prizes in life. Objectives are the smaller, intermediate steps that have to be taken to reach those goals.

Success on midterms and finals is an objective. It is an important, sometimes intimidating, but really quite *doable* step toward achieving your goals. Studying for—working toward—the midterm or final is *not* a bad thing, as long as you keep in mind the difference between objectives and goals. In fact, fixing your eye on the upcoming exam will help you to study more effectively. It gives you a more urgent purpose. It also gives you something specific to set your sights on.

Also, this book will help you study for your exams more effectively. By letting you see how knowledge may be applied—immediately and directly—to exams, it will help you acquire that knowledge more quickly, thoroughly, and certainly. Studying these exams will help you to focus your study to achieve success on the exams—that is, to help you attain the objectives that build toward your goals.

—*Alan Axelrod*

CONTRIBUTORS

Craig Blocker, *Professor of Physics, Brandeis University*

Robert I. Boughton, *Professor of Physics and Department Chairman, Bowling Green State University*

Gary Hastings, *Professor of Physics, Georgia State University*

Charles J. Montrose, *Associate Professor of Physics, Catholic University of America*

Lauren J. Novatne, *Teaching Assistant in Physics, California State University, Fresno*

Donald D. Pakey, *Associate Professor of Physics, Eastern Illinois University*

Ian K. Robinson, *Professor of Physics, University of Illinois*

Marc Sher, *Associate Professor of Physics, College of William and Mary*

Michael Strauss, *Professor of Physics, University of Oklahoma*

Christopher D. Wentworth, *Associate Professor of Physics, Doane College*

Jason Zimba, *Graduate Student Instructor in Physics, University of California, Berkeley*

ABOUT THE AUTHOR

Alan Axelrod, Ph.D., is the author of many books, including *Booklist* Editor's Choice *Art of the Golden West*, *The Penguin Dictionary of American Folklore*, and *The Macmillan Dictionary of Military Biography*. He lives in Atlanta, Georgia.

PREPARING YOURSELF

INTRODUCTION TO PHYSICS: WHAT THE PROFESSORS LOOK FOR

P HYSICS TAKES AS ITS SUBJECT THE STUDY OF *ALL* OBSERVABLE ASPECTS OF nature, defining "observable" from both an everyday perspective— "macroscopic" phenomena such as how various objects behave in motion—and a more arcane, but no less basic atomic and subatomic perspective: how molecules, atoms, and atomic particles behave. Physics also takes us to what we might call a "super" macroscopic perspective, to describe the structure and behavior of the universe on the very largest interstellar scales imaginable and at the very greatest velocities possible, near or at the speed of light. Some aspects of physics extend and quantify common sense and what we understand intuitively, whereas other aspects of physics seem stupefyingly to defy common sense, as is the case with relativity theory (Chapter 15). (Not to worry. Albert Einstein once defined common sense as "that layer of prejudice laid down prior to the age of sixteen." It doesn't pay to hang on too tightly to common sense in the realm of modern physics!)

> **The primary objective of the course is to provide a survey of physics at the college algebra level. For this, students should acquire and develop skills in problem solving and simple data evaluation as well as conceptual knowledge. The examinations are problem based in such a way that all the major concepts are covered. Data handling is covered in the laboratory portion of the course.**
>
> —Robert I. Boughton, Professor and Chairman of the Physics Department, Bowling Green State University

The scope of physics is so broad that it should offer something for everyone. But why kid ourselves? The fact is that physics *isn't* for everyone.

It makes certain demands on students, calling for:

◆ An imaginative willingness to strip reality to certain quantifiable essentials revolving around *forces*.

◆ A willingness and ability to think afresh about things (such as the motion of a ball or a pendulum) we ordinarily take for granted.

◆ A willingness and ability to work with abstract concepts and to visualize them.

◆ Competence with mathematics—at least on the level of algebra. Assuming a noncalculus introductory physics course, a nodding acquaintance with trigonometry helps, too.

These are the things that even an introductory course in physics calls for. You don't have to be a genius to excel in physics, but you do have to commit yourself to achieving *understanding* of the concepts that the physics course presents. Rote memorization of terms, theories, and formulas is not sufficient. You need to make a kind of contract with yourself to study, work practice problems, and ask questions to embrace each concept presented. For, above all else, physics requires the *self-discipline* to master concepts step-by-step, and to ensure that each step is thoroughly understood before moving on to the next.

This *step-by-step* approach cannot be emphasized strongly enough. It is fascinating to read the history of physics. What leaps out at you is the degree to which one concept relates to another and, indeed, is *built upon* another. Physicist A states a principle, on which physicist B builds, but, in building on it, creates a problem or raises a new question, which physicist C addresses and answers.

And so it goes.

When you study physics, you will find yourself following an analogous process. One concept provides the foundation for another. A new concept both answers and raises questions, generating the need for an additional concept. If you lose your grip, if you let a concept slip by, you will handicap your further understanding of the subject. The failures of understanding are relentlessly cumulative. Any engineer or architect is well aware of the consequences of failing to lay a sound foundation.

On the other hand, your commitment to step-by-step understanding will pay off. Some fields catalog huge amounts of facts, which may well be interesting, but physics attempts to distill such catalogs of facts into a few universal principles. When it succeeds—and when your efforts at understanding are rewarded—physics provides an

My first objective is that the students understand the basic laws of physics that are covered. This means that they not should just be able to write down equations, but that they should understand what the equation is actually saying and how to apply that equation to a problem. I am not interested in the students memorizing equations; they can always look them up if they need them later. I am more interested in their understanding the equations. For this reason, I always give the students the relevant equations on exams, so they don't have to spend time memorizing them.

A second objective is to give the students a sense that nature is described accurately by quantitative laws.

A third objective is to increase the problem-solving skills of the students, particularly for quantitative problems where the correct answer must be deduced by reasoning from one or more basic principles.

I want my students to gain an understanding of the basic laws of physics and be able to apply these laws to solve problems. The main obstacle to overcome here is the students' tendencies to jump to a conclusion without carefully thinking through and applying the relevant physics.

The exams that I give have problems to be solved that are similar to homework problems. These problems usually are not ones where the information given is simply plugged into a formula. Also, I do not give exams that directly test physics concepts (such as some multiple choice and essay questions). I have found that students often can quote the concepts fine, if they have been taught that way, but that they have difficulty applying these concepts to problems. I am interested in physics being a tool that the student can use to understand problems they encounter in the future studies and careers.

—Craig Blocker, Professor, Brandeis University

immensely satisfying "*aha!*" experience. It is truly a wide window on the world, opening out onto a very exciting view.

A Variety of Approaches

Some scientific fields—biology and psychology, for example—have a great many subdisciplines and specialties. Physics has its share of these, but it is among the more unified and compact of sciences, despite the broad scope of what it studies. Most physics professors are more or less agreed on the topics a "standard" introductory physics survey should address. They are the topics surveyed in Chapters 5 through 15 of this book. However, you will find variations in four areas:

1. Not all introductory physics courses are "standard." Some schools offer physics surveys that are specifically intended for nonscience majors. You may encounter courses with names like "Physics for Poets." Generally, these emphasize concepts and keep equations to a minimum. Other courses include a certain amount of math, but still put emphasis on concepts and themes. In this book, the exams in Chapter 20, from "Physics 101: Twentieth-Century Concepts of the Physical Universe," Catholic University of America, come from a thematic, concept-oriented course.

2. Some introductory courses focus on specific areas of physics— mechanics, for example, or electricity and magnetism—as a way of teaching basic physics principles.

3. Even in "standard" courses that survey all of the traditional physics areas, instructors differ significantly in the kinds of problems they expect students to master. It is not that the concepts presented differ, but the problems based on those concepts vary in complexity.

4. Be aware that physics, even on the survey level, may be offered in noncalculus and calculus flavors. Those who are comfortable with calculus know that it is a most efficient way of solving certain problems. However, few nonscience or nonmath majors come into introductory physics having mastered calculus; therefore, most introductory physics survey courses intended for nonmajors are noncalculus courses. The sample exams in this book are all from noncalculus physics courses.

To make certain that you are getting into a course that suits your needs and level of preparation, you should:

Students are well advised to understand key concepts first, and then to practice the application of these concepts with problem-solving skills. Good study habits include the repetition of exposure to material. The number of ways to be exposed (reading, lecture, homework, and lab activities) is as important as the number of times the student is exposed to the topics (reading the chapter two or three times).

—Lauren J. Novatne,
Teaching Assistant,
California State University, Fresno

The object of the course is to introduce students to two of the major developments in twentieth-century physics, namely, Einstein's theory of relativity and the quantum theory. Like Caesar's Gaul, the course is divided in three parts. The first part covers some classical ideas about motion from the Greeks through Newton and some elements of Maxwell's electromagnetic theory. The focus is on establishing the principle of relativity (the states of rest and of uniform motion are indistinguishable). The second part of the course deals with the theory of relativity, the postulates on which it is based (and their experimental basis), and a few of their consequences. The final portion of the course introduces some of the ideas of the quantum theory. In this portion the approach is quasi-historical, using the nineteenth and early twentieth century's scientists' struggles with understanding the nature of atomic structure as the motivation.

Students should develop their analytical abilities so that they are able to apply general scientific principles to analyze specific situations. In many instances, the analysis must be quantitative, although no heavy-duty mathematics is required.

—Charles J. Montrose,
Associate Professor,
Catholic University of America

◆ Read the course catalog description(s) carefully.

◆ Talk to the instructor before you sign up.

◆ Find out what kind of science background (if any) is required, expected, or considered helpful.

◆ Find out what kind of math competence is considered essential or helpful.

◆ If multiple courses are offered, work with an advisor or instructor to choose the approach that is right for you.

A Mathematical Question

Should you steer clear of physics if your math background is dim or your skills rusty?

There is no hard and fast answer.

The *fact* is that most physics courses do require a working knowledge of, at the least, algebra. If you hate math, you will probably hate physics. If math bewilders you, physics will not be fun, to say the least. However, just because you lack sufficient background or your skills are not up to par does not mean that you are disqualified from taking physics. Consider the following:

◆ What are your good reasons for taking physics? Apart from your being interested in the subject, do you need a physics course for a major requirement? (Most life sciences programs require some physics, as do engineering programs, premed programs, and the like.) Equally important, what role do you see for physics in your liberal education?

◆ You are a human being, which means that you are capable of learning; therefore, just because you lack certain math skills at the moment doesn't mean you can't acquire them. Talk to your advisor—and to a physics instructor—about remedial math, math review, or math tutoring.

◆ Look at it this way: If your math skills are inadequate to get you through introductory noncalculus physics, they are probably inadequate, period. Consider taking an appropriate math course—not just to prepare yourself for physics, but to fill in a serious gap in your education.

◆ If you lack confidence in your math and you don't want to take math courses, investigate any non-math-intensive physics courses that may be offered at your institution. Be aware that not all physics departments offer these physics-

As this is usually a student's first college physics course, one of my main objectives is to teach them to be rigorous, careful, and precise in their speech, calculations, and laboratory work. I also aim to teach them the models and concepts of classical mechanics, including Newton's Laws, momentum, and energy, and show them how these can be applied in a wide range of physical situations. Another objective is for the students to enjoy the process of reading a word problem, deciding how to solve it, and then successfully doing so. Finally, I attempt to prevent the compartmentalization of their physics knowledge: they should remember the material and be able to apply it even after the course is over.

Students need to develop the following skills and competencies:

1. Conceptual understanding of the concepts and laws of classical mechanics
2. General problem-solving ability, involving the greatest amount of independent thought possible
3. Both general and specific skills in the usage of scientific instruments
4. Theory and practice of error analysis
5. Mathematical methods, most importantly the theory and usage of the derivative and integral
6. Rigor and precision in thinking and problem-solving
7. Understanding and solving word problems

—Donald D. Pakey, Associate Professor, Eastern Illinois University

for-poets-type courses, courses that emphasize concepts over equations. You should also be aware that part of the "point" of physics is describing natural phenomena in mathematical terms. Yes, the basic concepts can be put into words, but most physics concepts are ultimately best expressed in the language of mathematics.

Great Expectations

Generally speaking, instructors and professors who teach introductory courses in various subjects expect great variation in the level of preparation among the students who come to them. Depending on your college or university, however, this may be somewhat less true among physics instructors. If physics is a prerequisite or requirement for many courses and programs in your institution, then, yes, wide variation is probably expected and taken into consideration in constructing coursework and exams. But, in a particular institution, if most students who elect introductory physics intend to major in science, the instructor most likely expects a reasonably sound high school level of preparation in science and in math.

Avoid unpleasant surprises. If you have doubts about the level of your preparation, talk it over with a physics advisor or instructor. Be honest and straightforward. Find out just what the instructor expects you to walk through the door with.

You should also discuss what is expected in class. Some instructors approach the survey course as a body of information to be absorbed. This is especially true in lecture-oriented courses. Others put more emphasis on lab work, outside reading, and independent problem solving. Whereas some instructors want you to memorize and master extensive amounts of prescribed materials, others are more interested in your learning basic principles, which you are then expected to apply to problems. Others still are most interested in your acquiring practical problem-solving skills.

It is the case, however, that, whatever else they look for, most scientists expect you to rely on your own initiative. In any science class, it helps to be a self-starter rather than someone who expects to be spoon-fed the course contents.

You should know that, at minimum, your instructor expects you to attend class, take lecture notes, read the textbook, do all assigned homework problems, and do practice problems on your own. In addition, you are expected to:

> The examinations, somewhat to the chagrin of a not insignificant portion of the students, cannot be completed either by rote learning of class notes and chapter summaries (and then regurgitating them on the exams) or by memorizing a list of formulas and plugging in numbers to solve exam problems. The exam questions require students to *understand* what the basic laws of physics mean, to *apply* general laws, both singly and in combination, to analyze and predict the outcomes of observations under specific conditions.
>
> There are no magic bullets. The road to success is an uphill one; otherwise cliches such as "pinnacle of success" would make no sense. You have to work at it. But you're not stupid, and if you work sensibly, you will succeed.
>
> Generally, don't let yourself get behind. Go to class. Once you're there, listen. Take *notes—don't* try to transcribe the lecture. Rewrite and annotate them. Keep up with the course on a class-by-class basis. Make sure that you understand Tuesday's class material before Thursday's class.
>
> Be a pest; ask questions of the instructor and TAs (if any) both in class and during their office hours. Enlist the aid of a TA in an "us-against-him (the instructor)" strategy.
>
> —Charles J. Montrose, Associate Professor, Catholic University of America

◆ Think problems through.

◆ Pay close attention to lab work.

◆ Study equations, diagrams, tables, and graphs as carefully as you study text.

◆ Ask questions.

Although, as mentioned, instructor expectations concerning your science background vary, at minimum you should have an understanding of the scientific method and its objectives and goals (see Chapter 4). You should also work to develop your powers of observation and your willingness to attend to detail. Moreover, you need a willingness to entertain new ideas and, while absorbing the details, you need to look beyond them to the "big picture."

Memorization and Study

Even for students who are fascinated by the physical sciences, introductory physics at the college level presents some formidable challenges. Every aspect of the science involves learning new concepts, new terms, and fundamental equations. Now, many instructors, in an effort to put the emphasis on concepts rather than formulas, do not demand that fundamental equations be committed to memory. On exams, they will include "formula sheets," which list the basic equations you'll need to solve the exam problems. However, you still need to know which formulas to choose from this "kit," and that means connecting concepts with appropriate equations.

◆ Commit yourself to memorizing terms, concepts, and fundamental equations.

◆ Associate a concept with the appropriate mathematical expression and vice versa. Don't try to learn a list of concepts or equations in a vacuum.

◆ Similarly, it is best to learn associated terms, concepts, and equations together. For example, the time to learn about entropy is when you are learning about the second law of thermodynamics.

◆ Identify and pay special attention to the *key* terms, concepts, and equations. Your instructor will point out many of these. Obviously, too, if a term, concept, or equation is used often in lecture or in your textbook, you must regard it as key, and you should become comfortable with it.

Keep up with the material; that is, do the readings and homework as assigned and don't fall behind. Secondly, I consider doing homework problems essential in learning physics. A student should make a serious attempt at all homework problems. If he or she has trouble with some of the problems, seek help from fellow students, teaching assistants, or myself. I also distribute solutions to the homework after it has been turned in. I encourage students to compare their solutions with my solutions and spend time understanding the differences.

—Craig Blocker, Professor, Brandeis University

The only way to learn physics is to do physics. The most important thing you can do to help yourself is to practice solving problems over and over. Just doing the homework isn't enough. Try to put each problem you see into a category: constant-acceleration, $F = ma$, energy conservation, momentum conservation, what have you. Then have a *standard approach* for each type of problem. The more practice problems you do, the more routine these standard strategies will seem.

Standard strategies work because introductory physics is not about smarts. It's about technique. Good technique will save you from making lots of mistakes. Draw clear force diagrams, make your sign conventions explicit, and so on. If you approach every problem differently, then you'll never see the big picture.

Practice, practice, practice. I am convinced that the more practice problems you do on your own time, the better you will do on exams.

—Jason Zimba, Graduate Student Instructor, University of California, Berkeley

> I don't flatter myself that they will remember how to solve circular motion problems two years after they take my course. I am really just trying to teach them how to analyze a problem logically, and how to formulate and carry out a strategy for its solution. Life is full of word problems, and physics is the science of word problems par excellence. Most of my students will not be using much physics after they leave my classroom, so I don't expect them to remember much of it.
>
> The course examinations are primarily a test of the students' problem-solving ability. They tend to reward those who have put in the most time and effort.
>
> —Jason Zimba, Graduate Student Instructor, University of California, Berkeley

◆ The best way to learn new terms, concepts, and equations is to *apply* them. Practice.

Practice. Physics doesn't just require brain power. It is not just a set of concepts. Much of the work of physics is calculation and manipulation of formulas. The more you practice applying the appropriate formulas to the appropriate problems, the easier day-to-day physics becomes.

Study Habits

In addition to practice—which is especially important in physics courses—the study habits that work for other college courses work well for physics, too. Because each lecture introduces new concepts, terms, and equations, it is very important to:

◆ Read textbook assignments *before* lectures.

◆ Keep up with your reading.

◆ Attend the lectures.

◆ Take good notes.

Visual Learning

Diagrams are very important in physics. Those that you will encounter in your textbook and in lecture presentations are more than decorative illustrations. They are important tools for understanding major concepts. A good way to study diagrams is to try drawing from memory at least the major features of key diagrams. At the

> Do the homework!!! Do several problems extra for every one assigned. If you don't understand it, ask the instructor for help. Too few students ask for help, get far behind, and run into trouble.
>
> —Marc Sher, Associate Professor, College of William and Mary

appropriate point in the course, you should practice drawing force diagrams (see Chapter 6) just as you practice working equations. An accurately drawn force diagram greatly simplifies many problems in mechanics.

Attitude

Don't let your experience with physics become a meaningless struggle with "exercises." Maintain an attitude of open-minded inquiry. Cultivate curiosity. Ask questions. Never think of a solved problem as the end, but, rather, as something with implications that open new avenues of inquiry. Keep thinking.

CHAPTER 2

KEYS TO SUCCESSFUL STUDY

THIS CHAPTER OUTLINES SKILLS THAT ARE INDISPENSABLE TO SUCCESSFUL STUDY, with special emphasis on skills important to the study of physics on the introductory level.

But let's not start thinking about physics just yet. Let's just start thinking. After all, isn't that what college and college courses are all about?

Well, not quite. *Think* about it.

For the ancient Greeks of Plato's day, from about 428 to 348 B.C., "higher education" really *was* all about thinking. Through dialogue, back and forth, the teacher and his student *thought* about physics, psychology, mathematics, the nature of reality—whatever. Perhaps the teacher evaluated the quality of his student's thought, but there is no evidence that Plato graded exams, let alone assigned the student a final grade for the course.

B.C. was a long time ago. Times have changed.

"Don't Study for the Grade!"

Today, you get graded. All the time, and on everything you do. Now, most of the professors, instructors, and teaching assistants from whom you take your courses will tell you that the real value of the course is in its contribution to your liberal education. A professor may even solemnly protest that he or she does *not* teach for the midterm and final. Nevertheless, the introductory, survey, and core courses almost always *include* at least one midterm and a final, and these almost always account for a very large part of the course grade. Even if the professor can disclaim with a straight face *teaching for* these exams, few students would deny *learning for* them.

The truth is this: More than anything else, most curriculum core courses *are* focused on the midterm and final exams.

"Grades Aren't Important"

Let's keep thinking. You and, most likely, your family are investing a great deal of time, cash, and sweat in your college education. It *would* be pretty silly if the payoff of all those resources were a letter grade and a numerical GPA. Ultimately, of course, the benefit is knowledge, a feeling of achievement, an intellectually and spiritually enriched life—*and* preparation for a satisfying and (you probably hope) financially rewarding career.

But, the fact is that if you don't perform well on midterms and finals, your path to all these forms of enrichment will be blocked. And, the fact *also* is that your performance is measured by grades. Sure, almost any reason you can think of for investing in college is more important than amassing a collection of A's and B's, but those stupid little letters are part of what it takes to get you to those other, far more important, goals.

"Don't Study *for* the Exam"

Most professors hate exams and hate grades. They believe that prodding students to pass tests and then evaluating their performance with a number or letter makes the whole process of education seem pretty trivial. Those professors who tell you that they "don't teach for the exam" may also advise you not to "study for the exam." That's not exactly what they mean. They want you to study, but to study to learn, not *for* the exam.

It's well-meaning advice, and it's true that if you study *for* the exam, intending to ace it, then promptly forget everything you've learned, you are making a pretty bad mistake. Yet, those same professors are part of a system that demands exams and grades, and if you don't study for the exam, the chances are very good that you won't make the grade and you won't achieve the higher goals you, your family, *and your professors* want for you.

Lose-Lose or Win-Win? Your Choice

When it comes to studying, especially in your introductory-level courses, you have some choices to make. You can decide grades are stupid, not study, and perform poorly on the exams. You can try not to study for the exams, but concentrate on higher goals, perform poorly on the exams, and never have the opportunity to reach those higher goals. You can study *for* the exams, ace them all, then flush the information from your memory banks, collect your A or B for the course, and move on without having learned a thing. These are all lose-lose scenarios, in which no one—neither you nor your teacher (nor your family, for that matter)—gets what he, she, or they really want.

Or you can go the win-win route.

We've used the words *goal* and *goals* several times. For an army general, winning the war is the goal, but to achieve this goal he must first accomplish certain *objectives*, such as winning battle number one, number two, number three, and so on. Objectives are intermediate goals or steps toward an ultimate goal.

Now, put exams in perspective. Performing well on an exam need not be an alternative to achieving higher goals, but should be an objective necessary to achieving those higher goals.

The win-win scenario goes like this: Use the fact of the exams as a way of focusing your study for the course. Focus on the exams as immediate objectives, crucial to achieving your ultimate goals. *Do* study for the exam, but not *for* the exam. *Don't* mistake the battle for the war, the objective for the goal; but *do* realize that you must attain the objectives to achieve the goal.

And *that*, ladies and gentlemen, is the purpose of this book:

◆ To help you ace the midterm and final exams in introductory-level physics courses . . .

◆ . . . without forgetting everything you learned after you've aced them.

This guide will help you *use* the exams to master the course material. This guide will help you make the grade—*and* actually learn something in the process.

Focus

How many times have you read a book, word for word, finished it, and closed the cover—only to realize that you've learned almost nothing from it? Unfortunately, it's something we all experience. It's not that the material is too difficult or that it's over our heads. It's that we mistake reading for studying.

For many of us, reading is a passive process. We scan page after page, the words go in, and, alas, the words seem to go out. The time, of course, goes by. We've *read* the book, but we've *retained* all too little.

Studying certainly involves reading, but reading and studying aren't one and the same activity. Or, we might put it this way: studying is intensely focused reading.

How do you *focus* reading?

Begin by setting objectives. Now, saying that your objective in reading a certain number of chapters in a textbook or reading your lecture notes is to learn the material is not very useful. It is an obvious but vague *goal* rather than a well-defined objective.

Why not let the approaching exam determine your objective?

"I will read and retain the stuff in chapters 10 through 20 because that's what's going to be on the exam, and I want to ace the exam."

Now, you at least have an objective. Accomplish this *objective*, and you will be on your way to achieving the *goal* of learning the material.

◆ An *objective* is an immediate target. A *goal* is for the long term.

Concentrate

To move from passive reading to active study requires, first of all, concentration. Setting up objectives (immediate targets) rather than looking toward goals (long-term targets) makes it much easier for you to concentrate. Few of us can (or would) put our personal lives completely on hold for four years of college, several years of graduate school, and *X* years in the working world to concentrate on achieving a goal. But, just about anyone can discipline himself or herself to set aside distractions for the time it takes to achieve the *objective* of studying for an exam.

Step 1. Find a quiet place to work.

Step 2. Clear your mind. Push everything aside for the few hours you spend each day studying.

Step 3. Don't daydream—*now*. Daydreaming, letting your imagination wander, is actually essential to real learning. But, right now, you have a specific objective to attain. This is not the time for daydreaming.

Step 4. Deal with your worries. Those pressing matters you can do something about *right now*, take care of. Those you can't do anything about, push aside for now.

Step 5. Don't *worry* about the exam. Take the exam seriously, but don't fret. Instead of worrying about the prospect of failure, use your time to eliminate failure as an option.

Plan

Let's go back to that general who knows the difference between objectives and goals. Chances are he or she also knows that you'd better not march off to battle without a plan. Remember, you don't want to *read*. You want to *study*. This requires focusing your work with a plan.

The first item to plan is your time.

Step 1. Dedicate a notebook or organizer-type date book to the purpose of recording your scheduling information.

Step 2. Record the following:

a. Class times

b. Assignment due dates

c. Exam dates

d. Extracurricular commitments

e. Study time

> **TIP:** If you are in doubt about what tasks to assign the highest priority, it is generally best to allot the most time to the most complex and difficult tasks and to get these done first.

Step 3. Inventory your various tasks. What do you have to do today, this week, this month, this semester?

Step 4. Prioritize your tasks. Everybody seems to be grading you. Now's *your* chance to grade the things they give you to do. Label high-priority tasks "A," middle-priority tasks "B," and lower-priority tasks "C." This will not only help you decide which things to do first, it will also aid you in deciding how much time to allot to each task.

Step 5. Enter your tasks in your scheduling notebook and assign order and duration to each according to its priority.

Step 6. Check off items as you complete them.

Step 7. Keep your scheduling book up to date. Reschedule whatever you do not complete.

Step 8. Don't be passive. Actively *monitor* your progress toward your objectives.

Step 9. Don't be passive. Arrange and rearrange your schedule to get the most time when you need it most.

Packing Your Time

Once you have found as much time as you can, pack it as tightly as you can.

Step 1. Assemble your study materials. Be sure you have all necessary textbooks and notes on hand. If you need access to library reference materials, study in the library. If you need access to reference materials on the Internet, make sure you're at a computer.

Step 2. Eliminate or reduce distractions.

Step 3. Become an efficient reader and note taker.

An Efficient Reader

Step 3 requires further discussion. Let's begin with the way you read.

TIP: Are you a—ugh!—*vocalizer?* A vocalizer is a reader who, during "silent" reading, either mouths the words or says them mentally. Vocalizing greatly slows reading, often reduces comprehension, and is just plain tiring. Work to overcome this habit— *except* when you are trying to memorize some specific piece of information. Many people do find it helpful to say over a sentence or two to memorize its content. Just bear in mind that this does not work for more than a sentence or paragraph of material.

Nothing has greater impact on the effectiveness of your studying than the speed and comprehension with which you read. If this statement prompts you to throw up your hands and wail, "I'm just not a fast reader," don't despair. You can learn to read faster and more efficiently.

Consider taking a "speed reading" course. Take one that your university offers or endorses. Most of the techniques taught in the major reading programs actually do work. Alternatively, do it yourself.

Step 1. When you sit down to read, try consciously to force your eyes to move across the page faster than normal.

Step 2. Always keep your eyes moving. Don't linger on any word.

Step 3. Take in as many words at a time as possible. Most slow readers aren't slow-witted. They've just been taught to read word by word. Fast and efficient readers, in contrast, have learned to read by taking in groups of words. Practice taking in blocks of words.

Step 4. Build on your skills. Each day, push yourself a little harder. Move your eyes across the page faster. Take in more words with each glance.

Step 5. Resist the strong temptation to fall back into your old habits. Keep pushing.

When you review material, consider skimming rather than reading. Hit the high points, lingering at places that give you trouble and skipping over the stuff you already know cold.

An Interactive Reader

TIP: The physical act of underlining actually helps you memorize material more effectively— though no one is quite sure why. Furthermore, underlining makes review skimming more efficient and effective.

Early in this chapter, we contrasted passive reading with active studying. A highly effective way to make the leap from passivity to activity is to become an *interactive* reader.

Step 1. Read with a pencil in your hand.

Step 2. Use your pencil to underline key concepts. Do this consistently. (That is, *always* read with a pencil in your hand.) Don't waste your time with a ruler; underscore freehand.

Step 3. Underline *only* the key concepts. If everything seems important to you, then up the ante and underline only the absolutely *most* important passages.

Step 4. If you prefer to highlight material with a transparent marker (a "Hi-Liter," for example), fine. But, you'll still need a pencil or pen nearby. Carry on a dialogue with your books by writing condensed notes in the margin.

Step 5. Put difficult concepts into your own words—right in the margin of the book. This is a great aid to understanding and memorization.

Step 6. Link one concept to another. If you read something that makes you think of something else related to it, make a note. The connection is almost certainly a valuable one.

Step 7. Comment on what you read.

TIP: Some instructors advise against using Hi-Liter-style markers to underscore books and notes, because doing so may discourage you from writing notes in the margin. If holding a Hi-Liter means that you won't also pick up a pencil to engage in a lively dialogue with books or notes, then it *is* best to lay aside the Hi-Liter and take the more *actively* interactive route.

Problems, Problems

Physics is both a conceptual and problem-related discipline. Even introductory-level physics courses involve a good deal of problem solving. Problem solving involves:

Step 1. Understanding the underlying concept(s)

Step 2. Recognizing what approaches and equations to plug into the concept(s)

Step 3. Applying the approaches and equations

Step 4. Calculating to solve the equations

The first two steps require intelligent study of your text and intelligent listening to lectures. The last two steps require *practice*. Application and calculation become easier—more automatic—with practice. This is analogous to learning the basics of music. Music history and music theory require study, but learning to play an instrument requires, above all else, practice.

TIP: Some students are reluctant to write in their textbooks, because it reduces resale value. True enough. But, is it worth an extra $5 at the end of the semester if you don't get the most out of your multithousand-dollar and multihundred-hour investment in the course?

Do your homework problems regularly. Do *extra* problems for practice. You cannot learn to solve problems by reading about them or by reading about the physical principles related to them. You learn to solve problems by actually working them.

Many physics concepts are challenging. You do *not* need the additional challenge of inadequate mathematical preparation. Physics is taught with and without calculus. Introductory-level physics courses are noncalculus courses. However, they draw heavily on algebra and at least touch on trigonometry. If you have any misgivings about the adequacy of your math skills:

TIP: Because physics problems can be difficult, it is best *not* to set yourself the goal of finishing a certain *number of problems* in any given study period. Instead, set an *amount of time* for the study period and solve as many problems as you can during that time.

◆ Talk to your instructor.

◆ Consider working with a math tutor recommended by your *physics* instructor.

◆ Dust off what math skills you have.

Taking Notes

The techniques of underlining, highlighting, paraphrasing, linking, and commenting on textbook material can also be applied to your classroom and lecture notes.

Of course, this assumes that you have taken notes. There are some students who claim that it is easier for them to listen to a lecture if they *avoid* taking notes. For a small minority, this may be true; but the vast majority of students find that note taking is essential when it comes time to study for midterms and finals. This does not mean that you should be a stenographer or court reporter, taking down each and every word. To the extent that it is possible for you to do so, absorb the lecture *in your mind,* then jot down major points, preferably in loose outline form.

Become sensitive to the major points of the lecture. Some lecturers will come right out and tell you, "The following three points are key." That's your cue to write them down. Other cues include:

◆ *Repetition.* If the lecturer repeats a point, write it down.

◆ *Excitement.* If the lecturer's voice picks up, if his or her face becomes suddenly animated, if, in other words, it is apparent that the material is at this point of particular interest to the speaker, your pencil should be in motion.

◆ *Verbal cues.* In addition to such verbal elbows in the rib as "this is important," most lecturers underscore transitions from one topic to another with phrases such as "Moving on to . . .," or "Next, we will . . .," or the like. This is your signal to write a new heading.

◆ *Slowing down.* If the lecturer gives deliberate verbal weight to a word, phrase, or passage, make a note of it.

◆ *Visual aids.* If the lecturer writes something on a blackboard, overhead projector, or in a computer-generated presentation, make a note.

◆ *Equations.* If the instructor presents an equation as key, believe it. Write it down carefully. (Most core-level physics instructors provide key equations on handouts. Many even include equation and formula sheets with the exams they give. However, don't depend on having these aids. Write down any formulas and equations presented in lecture.)

Filtering Notes

Some students take neat notes in outline form. Others take sprawling, scrawling notes that are almost impossible to read. Most students can profit from *filtering* the notes they take. Usually, this does *not* mean rewriting or retyping your notes. Many instructors agree that this is a waste of time. Instead, they advise, underscore the most important points, filtering out the excess.

If you have taken notes on a notebook or laptop computer, consider arranging the notes in clear outline form. If you have handwritten notes, however, it may not be worth the time it takes to create a neat outline. Instead, spend that time merely underlining or highlighting the most important concepts.

TIP: Although many instructors do not recommend rewriting or typing up your lecture notes, this may be useful in a subject like physics, which requires memorizing a good deal of conceptual and technical information, specialized vocabulary, and key mathematical expressions. You may want to give this tactic a try. If it helps, use it. However, if you find that rewriting or typing your notes does not help you, drop the practice.

Tape It?

Should you bring a tape recorder to class? The short answer is, probably not. To begin with, some instructors object to having their lectures recorded. Even more important, however, is the tendency to complacency that a tape recorder creates. You might feel that you don't have to listen very carefully to the "live" lecture, because you're getting it all on tape. This could be a mistake, because the live presentation is bound to make a greater impression on you, your mind, and your memory than a recorded replay.

So much for the majority view on tape recorders; however, a few professors actually *recommend* taping lectures. Fewer still suggest that students refrain from taking written notes in class and instead play back the tape later and take notes from that. If this method appeals to you, make sure of the following:

TIP: Before you tote your laptop or notebook computer to class, make certain that the instructor approves of such note-taking devices in the classroom. Most lecturers have no problem with these, but some find the tap-tap-tapping of maybe more than a hundred students distracting.

◆ Clear with your instructor the use of a tape recorder. Make sure he or she has no objections.

◆ Before you rely on the tape recorder, make sure that it works, that its microphone adequately picks up what's going on in the classroom, and that you are well positioned to record. The back of a large lecture hall is probably not a good place to be.

◆ Don't "zone out" or daydream during the lecture, figuring that you don't need to pay attention because you're getting it all on tape anyway. The value of the tape-recorded lecture is as a repetition and reinforcement of what you have already absorbed.

Remember: The mere fact that you have tape-recorded a lecture does not mean that you have learned the material. Technology is indeed marvelous, but it won't perform miracles. You have to listen to the tape and take notes from it.

◆ Make certain that you really do play back—and take notes from!—the taped lectures. This means setting aside the time to repeat the entire lecture.

Build Your Memory

Just as a variety of speed-reading courses are available, so a number of memory improvement courses, audio tapes, and books are on the market. It might be worth your while to scope some of these out, especially for a memory-intensive course like

TIP: Memorization is important, make no mistake, especially in an information-rich field like physics. However, it *is* usually overrated. Virtually all of the instructors and professors who have contributed to this book counsel students to *think* rather than merely memorize.

introductory psychology or if you are planning ultimately to go into a field that requires the memorization of a lot of facts. In the meantime, here are some suggestions for building your memory:

◆ Be aware that most so-called memory problems are really learning problems. You "forget" things you never *learned* in the first place. This is usually a result of passive reading or passive attendance at lectures—the familiar in-one-ear-and-out-the-other syndrome.

◆ Memorization is made easier by two seemingly opposite processes. First, we tend to remember information for which we have a context. It will indeed be hard to remember a bunch of definitions if you try to study these out of context—as a list rather than as parts of an interrelated system of thought.

◆ Second, memorization is often made easier if we break down complex material into a set of key phrases or words.

◆ It follows, then, that the best way to build your memory where a certain subject is concerned is to try to understand information in context. Get the "big picture."

◆ It also follows that, even if you have the big picture, you may want to break down key concepts into a few key words or phrases.

TIP: Many memory experts suggest that you try to put the key terms you identify into some sort of sentence; then memorize the sentence. Others suggest creating an acronym out of the initials of the key words or concepts. No one who lived through the Watergate scandal in the 1970s can forget that President Nixon's political campaign was run by CREEP (Committee to RE-Elect the President), and long after everything else taught in high school biology is forgotten, many students remember the sentence they memorized to learn how biologists classify organisms: Ken Put Candy On Fred's Green Sofa (kingdom, phylum, class, order, family, genus, species). Use whatever memory aids help *you*.

How We Forget

It is always better to keep up with class work and study than it is to fall behind and desperately struggle to catch up. This said, it is nevertheless true that most forgetting occurs within the first few days of exposure to new material. That is, if you "learn" 100 facts about Subject A on December 1, you may forget 20 of those facts by December 5 and another 10 by December 10, but by March 1 you may still remember 50 facts. The curve of your forgetting tends to flatten. Eventually, you may forget all 100 facts, but you will forget fewer and fewer each week.

Now, what does this mean to you?

It means that, midway through the course, you need to review material you learned earlier in the course. You cannot depend on having mastered it forever by having studied it two, three, four, or more weeks earlier.

The Virtues of Cramming

Ask any college instructor about last-minute cramming for an exam, and you'll almost certainly get a knee-jerk condemnation of the practice. But, maybe it's time to think beyond that knee jerk.

Let's get one thing absolutely straight. You cannot expect to pack a semester's worth of studying into a single all-nighter. It just isn't going to happen, especially in a physics class. However, cramming can be a valuable *supplement* to a semester of conscientious studying.

◆ You forget the most within the first few days of studying. (Or have you forgotten?)

Well, if you cram the night before the exam, those "first few days" won't fall between your studying and the exam, will they?

◆ Cramming creates a sense of urgency. It brings you face-to-face and toe-to-toe with your objective. Urgency concentrates the mind.

◆ Assuming you aren't totally exhausted, material you study within a few hours of going to bed at night is more readily retained than material studied earlier in the day.

Burning the midnight oil may not be such a bad idea.

> **TIP:** Any full-time college student studies several subjects each semester. This makes you vulnerable to interference—the possibility that learning material from one subject will interfere with learning material in another subject. Interference is usually at its worst when you are studying two similar or related subjects. If possible, arrange your study time so that work on similar subjects is separated by work on an unrelated one.

Cramming Cautions

Then again, staying up late before a big exam may not be such a hot idea, either. Don't do it if you have an early-morning exam. And, don't transform cramming into an all-nighter. You almost certainly need *some* sleep to perform competently on tomorrow's exam.

> **TIP:** If you *hate* cramming, don't do it. It's not for you, and it will probably only raise your anxiety level. Get some sleep instead.

Remember, too, that while cramming creates a sense of urgency, which may stimulate and energize your study efforts, it may also create a feeling of panic, and panic is never helpful.

Cramming is *not* a substitute for diligent study throughout the semester. But, just because you *have* studied diligently, don't shun cramming as a *supplement* to regular study, a valuable means of refreshing the mind and memory.

Polly Want a Cracker?

We've been talking a lot about memory and memorization. It's an important subject and, for just about any course of study, an important skill. Some subjects—physics included—are more fact-and-memory intensive than others. However, beware of relying too much on simple, brute memory. Try to assess what the professor really wants: Students who demonstrate on exams that they have absorbed the facts he or she and the textbooks have dished out? Or, students who demonstrate such skills as critical thinking, synthesis, analysis, and imagination? Depending on the professor's personal style and the kind of exam he or she gives (predominantly problem versus predominantly multiple choice, for example), you may actually be penalized for "parroting" lectures. ("I know what *I* think. I want to know what *you* think.")

Most physics instructors want students to be familiar with essential formulas and procedures, but they also value evidence of *understanding* that goes deeper than rote memorization.

Use *This* Book (and Get Old Exams)

One way to judge what the professor values and expects is to pay careful attention in class. Is discussion invited? Or, does the course go by the book and by the lecture? Also valuable are previous exams. Many professors keep these on file and allow students to browse through them freely. Fraternities, sororities, and formal as well as informal study groups sometimes maintain such files, too. These days, previous exams may even be posted on the university department's World Wide Web site. Most physics professors encourage students to pore over past exams and mine them for practice problems.

Of course, you are holding in your hand a book chock-full of sample and model midterm and final exams. Read them. Study them. And let them focus your study and review of the course.

Study Groups: The Pro and The Con

In the old days (whenever that was), it was believed that teachers *taught* and students *learned*. More recently, educators have begun to wonder whether it is possible to *teach* at all. A student, they say, *learns* by teaching himself or herself. The so-called teacher (who might better be called a learning facilitator) helps the student teach himself or herself.

Well, maybe this is all a matter of semantics. Is there really a difference between *teaching* and *facilitating learning*? And between *learning* and *teaching oneself*? The more important point is that the focus in education has turned away from the teacher to the student, and students, in turn, have often responded by organizing study groups, in which they help each other study and learn.

These can be very useful:

◆ In the so-called real world (that is, the world after college), most problems are solved by teams rather than individuals.

◆ Many people come to an understanding of a subject through dialogue and question and answer.

◆ Studying in a group (or even with one partner) makes it possible to drill and quiz one another.

◆ In a group, complex subjects can be broken up and divided among members of the group. Each one becomes a "specialist" on some aspect of the subject, then shares his or her knowledge with the others.

◆ Studying in a group may improve concentration.

◆ Studying in a group may reduce anxiety.

Not that study groups are without their drawbacks:

◆ All too often, study groups become social gatherings, full of distraction (however pleasant) rather than study. This is the greatest pitfall of a study group.

◆ All members of the group must be highly motivated to study. If not, the group will become a distraction rather than an aid to study, and it is also likely that friction will develop among the members, some of whom will feel burdened by "freeloaders."

◆ The members of the study group must not only be committed to study, but to one another. Study groups fall apart—bitterly—if members, out of a sense of competition, begin to withhold information from one another. This *must* be a Three Musketeers deal—all for one and one for all—or it is worse than useless.

◆ The group may promote excellence—or it may agree on mediocrity. If the latter occurs, the group will become destructive.

In summary, study groups tend to bring out the members' best, as well as worst, study habits. It takes individual and collective discipline to remain focused on the task at hand, to remain committed and helpful to one another, to insist that everyone shoulders their fair share, and to insist on excellence of achievement as the only acceptable standard—or, at least, the only valid reason for continuing the study group.

CHAPTER 3

SECRETS OF SUCCESSFUL TEST TAKING

SOMETIMES IT SEEMS THAT THE DIFFERENCE BETWEEN ACADEMIC SUCCESS AND something less than success is not smarts versus nonsmarts or even study versus nonstudy, but simply whether or not a person is good at taking tests. That phrase—"good at taking tests"—was probably first heard back when the University of Bologna opened for business late in the eleventh century. The problem with phrases like this is that they are true enough, yet not very helpful.

Fact: Some people *are* and some people *are not* good at taking tests.

So what? Even if successful test taking doesn't come to you easily or naturally, you *can* improve your test-taking skills. Now, if you happen to have a knack for taking tests, well, congratulations! But, that won't help you much if you neglect the kind of preparation discussed in the previous chapter.

Why Failure?

In analyzing performance on most tasks, it is generally better to begin by asking what you can do to succeed. But, in the case of taking tests, success is largely a question of *avoiding* failure. So let's begin there.

When the celebrated bank robber Willy "The Actor" Sutton was caught, a reporter asked him why he robbed banks. "Because that's where they keep the money," the handcuffed thief replied. At least one answer to the question of why some students perform poorly on exams is just as simple: "Because they don't know the answers."

There is no "magic bullet" in test taking. But, the closest thing to one is *knowing the course material cold.* Pay attention, keep up with reading and other assignments, attend class, listen in class, take effective notes—in short, follow the recommendations of the previous chapter—and you will have taken the most important step toward exam success.

Yet, have you ever gotten your graded exam back, with disappointing results, read it over, and found question after question you now realize you *could* have answered correctly?

"I *knew* that!" you exclaim, smacking yourself in the forehead.

What happened? You *really* did prepare. You *really* did know the material. What happened?

Anxiety Good and Bad

The great American philosopher and psychologist William James (1842–1910) once advised his Harvard students that an "ounce of good nervous tone in an examination is worth many pounds of . . . study." By "good nervous tone," James meant something very like anxiety. You *should not* expect to feel relaxed just before or during an exam. Anxiety is natural.

> **TIP:** Most midterm and final exams really *are* representative of the course. If you have mastered the course material, you will almost certainly be prepared to perform well on the examination. Very few instructors purposely create deceptive exams or trick questions or even questions that require you to think beyond the course. Most instructors are interested in creating exams that help you and them evaluate your level of understanding of the course material. Think of the exams as natural, logical features of the course, not as sadistic assignments designed to trip you up. Remember, your success on an examination is also a measure of the instructor's success in presenting complex information. Very few teachers can—or want to—build careers on trying to *fail* their students.

Anxiety is natural because it is helpful. Good nervous tone, alert senses, sharpened perception, adrenaline-fueled readiness for action are *natural* and *healthy* responses to demanding or threatening situations. We are animals, and these are reactions we share with other animals. The mongoose that relaxes when it confronts a cobra is a dead mongoose. The student who takes it easy during the physics final . . . Well, the point is neither to fight anxiety nor to fear it. Accept it, and even welcome it as an ally. Unlike our hominid ancestors of distant prehistory, we no longer need the biological equipment of anxiety to help us fight or fly from the snapping saber teeth of some animal of prey, but every day we do face challenges to our success. Midterms and finals are just such challenges, and the anxiety they provoke is real, natural, and unavoidable. It may even help us excel.

What good can anxiety do?

◆ Anxiety can focus our concentration. It can keep the mind from wandering. This makes thought easier, faster, and, often, more acute and effective.

◆ Anxiety can energize us. We've all heard stories about a 105-pound mother who is able to lift the wreckage of an automobile to free her trapped child. This isn't fantasy. It really happens. And, just as adrenaline can provide the strength we need when we need it most, it can enhance our ability to think under pressure.

◆ Anxiety moves us along. Anxious, we work faster than when we are relaxed. This is valuable because, in most midterms and finals, limited time is part of the test.

◆ Anxiety prompts us to take risks. We've all been in classes in which the instructor has a terrible time trying to get students to speak, discuss, and venture an opinion. "Come on, come on," the poor prof protests, "this is like pulling teeth!" Yet, when exam time comes, all heads are bent over bluebooks or answer sheets, and the answers—*some* kind of answers—are flowing forth or, at least, grinding out. Why? Because the anxiety of the exam situation overpowers the inertia that keeps most of us silent most of the time. We take the risks we have to take. We answer the questions.

◆ Anxiety can make us more creative. This is related to risk taking. "Necessity," the old saying goes, "is the mother of invention." Phrased another way: *we do what we have to do.* Under pressure, many students find themselves taking fresh and creative approaches to problems.

So, don't shun anxiety. But, unfortunately, the scoop on anxiety isn't all good news, either.

Anxiety evolved as a mechanism of *physical* survival. Biologists and psychologists talk of the fight-or-flight response. Anxiety prepares a threatened animal either to fight the threat or to flee from it. The action is physical and, typically, very short term. In our civilized age, the threats are generally less physical than intellectual and emotional, and they tend to be of longer duration than a physical fight or a physical flight. This means that the anxiety mechanism does not always work to enhance our chances for survival or, at least, our chances to survive the course by performing well on exams. Some of us are better than others at adapting the *physical* benefits of anxiety to the *intellectual* and *emotional* challenges of an exam. Some of us, unfortunately, are unable to benefit from anxiety, and, for still others of us, anxiety is downright harmful. Here are some of the negative effects anxiety may have on exam performance:

◆ Anxiety can make it difficult to concentrate. True, anxiety focuses concentration. But, if it focuses concentration on the anxious feelings themselves, you will have less focus left over for the exam. Similarly, anxiety may cause you to focus unduly on the perceived consequences of failure.

◆ Anxiety causes carelessness. If anxiety can prompt you to take creative risks, it can also cause you to rush through material and, therefore, to make careless mistakes or simply to fail to think through a problem or question.

◆ Anxiety distorts focus. Anxiety may impede your judgment, causing you to give disproportionate weight to relatively unimportant matters. For example, you may become fixated on solving a lesser problem at the expense of a more important one. This is related to the next point.

♦ Anxiety may distort your perception of time. You may think you have more or less of it than you really do. The result may be too much time spent on a minor question at the expense of a major one.

♦ Anxiety tends to be cumulative. Many test takers have trouble with a question early in the exam, then devote the rest of the exam to worrying about it instead of concentrating on the *rest of the exam.*

> **TIP:** Exam questions are battles in a war. No general expects to win every battle. Accept your losses and move on. Dwell on your losses, and you will continue to lose. Note that in most physics problems partial credit is given for whatever you do manage to accomplish.

♦ Anxiety drains energy. For short periods of time, anxiety can be energizing and invigorating. But, if anxiety becomes chronic, it begins to tire you out. You do not perform as well.

♦ Anxiety can keep you from getting the rest you need. If it is generally unwise to stay up all night *studying* for an exam, how much less wise is it to stay up uselessly *worrying* about one?

How can you combat anxiety?

Step 1. *Don't* fight it. Accept it. Remember, anxiety is a *natural* response to a stressful situation. Remember, too, that some degree of anxiety aids performance. Try to learn to accept anxiety and *use* it. Let it sharpen your wits and stoke the fires of your creativity.

Step 2. Don't worry about how you feel. Focus on the task. Usually, you will feel better once you overcome the initial jitters and inertia. William James, who lauded "good nervous tone," also once observed that we do not run because we are frightened, but we are frightened because we run. If you concentrate on your fear and act as if you are afraid, you will become even more fearful.

> **TIP:** More serious stimulant drugs ("speed," "uppers") are *never* a good idea. They are both illegal and dangerous (possibly even deadly), and, for that matter, their effect on exam performance is unpredictable. The chances are they will impede performance rather than aid it (though you may erroneously *feel* that you are doing well). Also avoid over-the-counter stimulants. These are caffeine pills and will probably increase anxiety rather than improve performance.

Step 3. Prepare for the exam. Do whatever you must do to master the material. Build confidence in your understanding of the course, and your anxiety should be reduced.

Step 4. Get a good night's sleep before the exam.

Step 5. Avoid coffee and other stimulants. Caffeine tends to increase anxiety. (However, if you are a caffeine fiend, don't pick the day or two before a big exam to kick the habit. You *will* suffer withdrawal symptoms.)

Step 6. Try to get fresh air shortly before the exam. This is especially valuable if you have been cooped up for a long period of study. Take a walk. Get a look at the wider world for a few minutes.

Have a Plan, Make a Plan

A large component of destructive anxiety—probably the largest—is fear of the unknown. Reduce anxiety by taking steps to reduce the component of the unknown.

Step 1. To repeat—Do whatever is necessary to master the material on which you will be examined.

Step 2. Use the exams in this book to familiarize yourself with the kinds of exams you are likely to encounter.

Step 3. If possible, examine old exams actually given in the course.

Let's pause here before going on to Step 4. Just reading over the exams in this book or leafing through exams formerly given in the course will not help you much. Analyze:

◆ The types of questions asked. Are they concept- and information-oriented, or are they problems? Do the questions call for regurgitation of memorized material, or are they more think-oriented, requiring significant initiative to answer?

◆ Don't just predict which questions you could and could not answer or which problems you could or could not solve. Try actually answering some of the questions and solving some of the problems.

◆ If you are looking at sample exams with answers, evaluate the answers. How would you grade them? What would you do better?

◆ Don't just sit there, *do* something. If your analysis of the sample exams or old exams reveals areas in which you are weak, address those weaknesses.

An effective way to reduce the unknown is to create a plan for confronting it. Let's go on to Step 4.

Step 4. Make a plan. Begin *before* the exam. Decide what areas you need to study hardest. Based on your textbook notes and—especially—on your lecture notes, try to anticipate what kinds of questions will be asked. Work up answers or sketches of answers for these.

Step 5. Make sure you've done the simple things. The night before your exam, make certain that you have whatever equipment you'll need. If you will be allowed to use reference materials, bring them. If you are permitted to write on a laptop or notebook computer, make certain your batteries are fully charged. Same goes for your calculator.

TIP: Don't make the mistake of devoting all of your time to trying to make *last-minute* repairs to weak spots ("I've got one hour to read that textbook I should have been reading all along!") only to ignore your strengths. Develop your strengths. With any luck at all, the exam will give you an opportunity to show yourself at your best—not just trip you up at your worst. Be as prepared as you can be, but, remember, there is nothing wrong with excelling in a particular area. Play to your strengths, not your weaknesses.

TIP: Do plenty of practice problems with two objectives in mind: The first is simply to get better and more efficient at working the problems. The second is to uncover and diagnose your weak spots. Discover the gaps you need to fill—the kind of problems you need to work harder on.

If you are writing the exam longhand, make certain you have pens, pencils, paper. Bring a watch.

Step 6. Expect a shock. The first sight of the exam usually packs a jolt. At first sight, questions may draw a blank from you. Questions you were sure would appear on the exam will be absent, and some you never expected will be staring you in the face. Don't panic. *Everybody feels this way.*

Step 7. Write nothing yet. Read through the exam thoroughly. Be certain that you (1) understand any instructions and (2) understand the questions.

Step 8. If you are given a choice of which questions to answer, choose them now. Unless the questions vary in point value assigned to them, choose those that you feel most confident about answering. Don't challenge yourself.

Alternative Step 8. If you are required to answer all the questions, identify those about which you feel most confident. Answer these first.

> **TIP:** When you study for an exam, it is usually best to assign high priority to the most complex and difficult issues, devote ample time to these, and master them first. When you take an exam, however, and you are under time pressure, tackle first what you can most readily and thoroughly answer, then go on to more doubtful tasks. Your professor will be more favorably impressed by good answers than by failed attempts to answer questions you find difficult.

Step 9. After you have surveyed the exam, create a time budget: note—jot down—how much time you should give to each major question or problem.

Step 10. Reread the question or problem before you begin to write. Then plan your answer or solution.

Plan Your Solution

Here is where practice really helps! If playing the piano required relearning scales and chords *each* time one put his or her fingers to the keys, there would be very few pianists in the world. But, of course, experienced pianists approach their instrument with a well-rehearsed knowledge of scales and chords. With this background, built by practice, even a brand-new score looks at least somewhat familiar.

If you have read your text, listened in lecture, and, most of all, practiced problem solving, the problems you see on the exam will, likewise, look familiar. Indeed, most physics professors purposely put modified versions of homework problems on their exams.

Begin, then, by surveying the problems. Identify those that are *most* familiar. Work on those first—and use on them the procedures that you have practiced.

Some problems are relatively easy. Some are tough. Many physics problems, however, are not so much difficult as they are complex. That is, they are made up of a number of subproblems. With practice, you should be able to recognize how these complex problems break down into subproblems. Decide what subproblems are

TIP: Reduce the number of variables you carry into the exam. Practice the basic mathematical operations you know you will need. Be very certain that you are fully familiar with the advanced functions of your calculator. You'll have your hands full doing the physics in the exam. You can't afford taking time to learn the math you should have at your fingertips.

contained within the main problem. Decide how you will solve them. This is a very important step in planning your solution.

Plan Your Answer

Although most physics exams are heavily weighted toward problem solving, many introductory courses also call for answering conceptual or information-related questions.

Perhaps you have heard a teacher or professor comment on the exam he or she has just handed out: "The answers are in the questions." This kind of remark is more helpful than it may at first seem.

Begin by looking for the key words in the question. These are the verbs that *tell* you what to do, and they typically include:

- Compare
- Contrast
- Criticize
- Define
- Describe
- Discuss
- Evaluate
- Explain
- Illustrate
- Interpret
- Justify
- Outline
- Relate
- Review
- State
- Summarize
- Trace

Of these key words, the following are most often found in the kind of short-answer questions you are likely to encounter in the conceptual portion of a physics exam:

◆ To **define** something is to state the precise meaning of the word, phrase, or concept. Be succinct and clear.

◆ To **illustrate** is to provide a specific, concrete example.

◆ To **outline** is to provide the main features or general principles of a subject. This need not be in paragraph or essay form. Often, outline format is expected.

◆ To **state** is similar to *define*, though a statement may be even briefer and usually involves delivering up something that has been committed to memory.

◆ To **summarize** is to state briefly—in sentence form—the major points of an argument, position, or concept; the principal features of an event; or the main events of a period.

Approaching the Short-Answer Test

The short-answer exams you may get in physics courses are of two major kinds:

1. **Recall exams** include questions that call for a single short answer (usually there is a single correct answer) and fill-in-the-blank questions, in which you are asked to supply missing information in a statement or sentence.

2. **Recognition exams** include multiple-choice tests, true-false tests, and matching tests (match one from column A with one from column B).

If the exam is a long one and time is short, invest a few minutes in surveying the questions, so that you can be certain to answer those you are confident of, even if they come near the end of the exam.

Be prepared to answer multiple-choice questions through a process of elimination, if necessary. Usually, even if you are uncertain of the one correct answer among a choice of five, you *will* be able to eliminate one, two, or three answers you know are *incorrect*. This at least increases your odds of giving a correct response.

Unless your instructor has specifically informed you that he or she is penalizing guesses (actually taking points away for incorrect responses versus awarding zero points to unanswered questions), *do* guess the answers even to those questions that leave you in the dark.

> **CAUTION! Many physics courses do include multiple-choice exams—that is, exams in which the right answer is included among choices presented to you. However, in many cases, providing the answer is not enough to earn full credit for the item. If you are required to show your work—to show how you arrived at the answer—be sure to do so. Obviously, having to show your work makes guessing a less viable option.**

Plan your responses to true-false questions carefully. Look for telltale qualifying words, such as *all, always, never, no, none,* or *in all cases.* Questions with such absolute qualifiers *often* require an answer of *false,* because relatively few general statements are always either true or false. Conversely, questions containing such qualifiers as *sometimes, usually, often,* and the like are frequently answered correctly with a response of *true.*

A final word on guessing: *First guess, best guess.* Statistical evidence consistently shows that a first guess is more likely to be right than a later one. Obviously, if you have responded one way to a question and then the correct answer suddenly dawns on you, *do* change your response. But, if you can choose only from a variety of guesses, go with your first, or gut, response.

Take Your Time

Yes, yes, yes, this is easier said than done. But, the point is this:

◆ Plan your time.

◆ Work efficiently, but not in a panic.

◆ Make certain that your responses are legible. This is especially important in calculations and equations.

◆ Take time to double-check your math.

◆ Be certain that you include the appropriate units [kg, J (joules), m (meters), s (seconds), etc.] in your response. Points are lost for failing to use the proper units.

◆ In word responses, take time to spell correctly. Even if an instructor does not consciously deduct points for misspelling, such basic errors will negatively influence the evaluation of the exam. Correct spelling of physics terms is especially important—and can be difficult. Be careful.

◆ In responses that require a sketch or a graph, make sure what you draw is legible, and make doubly sure that you have labeled all features correctly and clearly.

◆ If a short answer is called for, make it short. Don't ramble.

◆ Use *all* of the time allowed. The instructor will *not* be impressed by a demonstration that you have finished early. If you have extra time, reread the exam. Look for careless errors. Do not, however, heap new guesses on top of old ones; where you have guessed, stick with your first guess.

CHAPTER 4

THINKING LIKE A PHYSICIST: KEYS TO UNDERSTANDING PHYSICS

HOW DOES A PHYSICIST THINK? LET'S ASK FIRST: WHAT IS PHYSICS?

The answer is a bit daunting, because physics is concerned with *all* observable aspects of nature. Is that broad enough for you?

Well, it gets even broader: Physicists define *observable* on a macroscopic (large-scale), as well as a submicroscopic (atomic and subatomic), level. In short, physics deals with the structure of matter and with the interactions between the fundamental constituents of the observable universe. The goal of this study, ultimately, is to formulate a comprehensive set of principles that brings together and explains all the apparently disparate and diverse phenomena that surround us. (That goal has yet to be reached.)

So *this* is what a physicist is supposed to think about? Just where are you supposed to dig in?

The answer is that you begin at the most basic level of what is observed. Physicists engage the world around them by attempting to describe and explain the physical changes in objects. Typically, this means reducing phenomena to idealized mental models of reality, which—again, typically—are expressed in mathematical terms as equations. Although these equations are derived from idealized conditions, they (typically) approximate reality closely enough to allow prediction of the behavior of even very complex systems, including those on a subatomic level, which we cannot observe directly.

Although the scope of physics is as broad as the universe itself, it approaches this field from the limited perspective of very basic *physical* quantities:

◆ Length

◆ Time

◆ Mass

◆ Velocity

◆ Acceleration

◆ Electric current

◆ Temperature

◆ Amount (of a substance)

◆ Luminous intensity

Physicists attempt to express phenomena in the observable universe in terms of these dimensions with the object of understanding the relationships among phenomena. Most often, these relationships are described in terms of *forces*.

So, how do you think like a physicist?

You acquire the habit of close observation (basic to any science) and to thinking about even everyday phenomena (*especially* everyday phenomena) in quantitative terms—that is, in terms of the quantities just mentioned and in terms of relationships (especially cause-and-effect relationships) among phenomena. Think about how everything you see, feel, or hear can be described in these quantities and as the effect of some identifiable and measurable cause. Think of these things as the product and evidence of changes in energy and matter. Get used to straddling the everyday world of real objects in motion, of electric currents, of heat transfer, and the mathematical world that expresses—within the mind—these worldly happenings.

The Ground Rules

All of the foregoing might be summarized in a single sentence: *A physicist thinks like a scientist.*

We all live in the universe, and we all have experiences with motion, heat, electricity, and light. Consciously, as well as unconsciously, we continually form opinions about the phenomena around us and the nature of reality. Scientists, including physicists, try to get beyond opinion based on such things as intuition and common sense. To do this, they employ the *scientific method.*

Scientific Method

The scientific method is central to all scientific inquiry. As different as, say, physics is from psychology, both the physicist and the psychologist use the scientific method to guide their investigations.

The *scientific* method—this must be a most profound and complex method, indeed!

Well, not really.

The scientific method is nothing more or less than a systematic approach to a common-sense way of looking at the world—with the emphasis on *systematic*. It works like this:

1. A problem is stated.
2. Facts relating to the problem are gathered.
3. A solution to the problem is proposed: a hypothesis.
4. The hypothesis is tested.

Let's look more closely at these four steps. As the elements of the scientific method, they are the most important early steps you can take toward thinking like a physicist.

The first step is to formulate a problem. For physicists, the problem is typically about explaining the change in some object or its behavior. Like most scientists, physicists are more interested in the mechanisms by which the phenomena they study operate than in questions of the ultimate purpose of such phenomena. The ultimate goal toward which physics is directed is a comprehensive model of the universe, a set of mathematical expressions and principles from which an understanding of all relationships among phenomena may be understood. But, even this goal does not speculate about the *purpose* of it all. Such questions are in the realm of philosophers and theologians, not scientists.

Anyone—not just a scientist—can raise questions about the mechanisms of the physical world. The fact is, however, that most of us don't, at least not consistently and systematically. Scientists may or may not be *born* curious, but it's a sure bet that they all *develop* curiosity. Without it, Step 1 is impossible.

You will increase your chances of excelling in physics—and enjoying the course—if you work at cultivating your curiosity about the world around you. Come to class and approach your textbooks with an attitude of curiosity, and learning will not only be easier, but much more meaningful.

After formulating a problem to study, the next step in the scientific method is to gather facts (data). Facts are gathered by

◆ Firsthand observation

◆ Measurement

◆ Counts

◆ Review of records of past observations

This material must be evaluated for reliability (Is it reliable? Why? Why not? *How* reliable is it?) and filtered for relevance to the problem. These processes typically require observational skill, an eye for detail, patience, a willingness to work hard, and objectivity. The latter quality is especially important. Data must be approached without prejudice. To prejudge the data is to compromise its usefulness.

Next, an educated guess is formulated concerning the relation between the problem identified and the facts gathered. This is the *hypothesis*. Its very special quality is that it *must* be falsifiable—that is, a statement capable of being proved false. It is *not* necessary that a hypothesis be proved true. Many, many common scientific hypotheses cannot be proved true beyond a doubt, yet they are still useful—as long as they cannot be proven beyond a doubt false.

The next step in the scientific method has as its object, therefore, not necessarily proving the hypothesis true, but proving it false. This is called *testing the hypothesis*. The scientist does his or her utmost to find ways to contradict or disprove the hypothesis. This is the phase of the scientific method that many people—not just in college, but throughout history—have the most trouble with. It is human nature to hang on to cherished beliefs, even in the face of evidence to the contrary. In the seventeenth century, when the astronomer Galileo introduced observational evidence supporting Copernicus's idea that the earth is not the center of the universe, he was put on trial by the Church, and came perilously close to losing his life. Similarly, ever since Charles Darwin developed the theory of evolution in the mid-nineteenth century, there has been no shortage of individuals and religious organizations who have opposed it.

Why is the theory of evolution, for example, called a *theory* rather than a hypothesis? A hypothesis that is repeatedly tested and never falsified is raised to the status of theory. Yet, note that the theory of evolution has not been proved true. It has simply never been proved false. It has stood as a sufficient explanation of countless particular instances. It has never been successfully contradicted.

Testing of a hypothesis can be through experiment. An experiment is a test in which the researcher controls as many of the variables—the context and inputs of the test—as possible. Ideally, the effect of altering only one variable at a time is observed and noted.

Where an experiment is not possible or practical, *observation* is used. Darwin, for example, did not try to experiment with evolution in a laboratory, but made meticulous observations of the fauna of South America and Africa. To test his theory of general relativity, Albert Einstein proposed three observable phenomena that could be explained by the theory: one involving the orbit of Mercury, another the deflection of starlight passing near the sun caused by the warping of space near the solar mass, and, finally, the gravitational redshift of light as a demonstration of space-time. (These concepts will be explained in Chapter 15.)

In cases in which it is impractical or impossible to gather facts by experiment or even by direct observation, physicists use a method called a *thought experiment.* This is a systematic hypothetical or imaginary simulation of reality. For example, neither Albert Einstein nor anyone else can travel at or near the speed of light and report the effects of such travel. But, this hasn't stopped physicists from constructing thought experiments to investigate the effects of such high-speed travel. You will encounter examples of thought experiments in Chapter 15.

Watch Your Language

Just as you must get accustomed to using the scientific method in your approach to physics, keeping in mind its four basic steps, so you must understand that physics has a vocabulary all its own. Essential to success in a physics course is learning the vocabulary and then using it *accurately.* This requires memorization, but it also requires care and thought. Typically, physics terms have limited and precise definitions. In contrast to everyday language, nuance and shades of meaning are avoided. Objective description and labeling are sought.

It is a good idea to create a running list of the technical and specialized words you encounter, complete with definitions *and* examples of *usage.* To be sure, your textbook will have a glossary of terms, and you will also find a basic glossary in this book. But, such resources are no substitute for your actually taking the time and effort to create your own list. Don't strive for completeness, but do make note of the terms you most frequently encounter, the terms that are obviously key.

Although you should use the language of physics with thought and care, you don't have to become paralyzed by fear of misusing terms. Just make certain that you understand each new term you encounter. Don't let any slip by. Such slips are cumulative. If you fail to understand term A, term B becomes that much more difficult to grasp. By the time you arrive at term C, you may be totally in the dark. It is especially important to understand the correct terms for units of measure of energy, motion, mass, temperature, and so on.

Remember, too, that the first step toward study of any physics problem or situation is *description,* and physics provides an ample and precisely defined vocabulary to facilitate such description. The specialized terms of this science are the *tools* of the discipline. To work with physics, even at the introductory level, you must have a sound knowledge of the basic tools. It's pretty hard to drive a nail with a screwdriver. Pick up the hammer instead. It's pretty hard to talk about the subject of *mass* if you think the word means the same thing as *weight.*

Another List

In addition to your running list of specialized terms, you should compile a list of the major assumptions, theories, and basic equations and formulas that you frequently encounter or that are pointed out to you as especially important. Again, you will certainly find these in your textbook and hear them in lecture, but writing them down for yourself is a great, *active* way of learning them.

Thinking Graphically

Many principles and experimental findings in physics are not stated in words and equations only, but graphically, either as actual graphs or as sketches (such as *free-body diagrams*). Include these in your custom-compiled list. Don't overlook graphs and illustrations. Work at interpreting and understanding them. Exams often include visual material, and you may even be asked to create drawings to illustrate some principle or phenomenon.

TIP: Think of this list as an annotated index. You should not only write down the particular law, principle, or equation, but make a note of where in your textbook it is discussed and explained.

As for graphs, be sure to *study* them—and by *study*, we mean *interpret*: think about the meaning of what you are seeing. The best way to go about this is to look at a graph and try to explain *in words* the relationships illustrated. This is not something you learn to do all at once. It takes practice—just as reading words on a page or notes on a musical score takes practice. There is nothing about C-A-T that *looks* like a cat, but, with practice, we learn to see a cat when we read *cat*. The same principle applies when working with graphs.

Doing the Numbers

Some introductory physics courses designed for nonscience majors bend over backward to avoid equations. Even such courses, however, cannot avoid them all. And, the fact is that most physics courses, at whatever level, use equations as the principal language of the discipline.

In studying physics, there is no substitute for a sound mathematics background. If your math is rusty, polish it up. But, it is also helpful to approach equations the same way you approach graphs: Get into the habit of mentally translating equations into words. Don't just regard them as abstract sets of symbols. They describe important *physical* processes. They are not purely mathematical expressions.

Experimental Work

Some introductory physics courses are taught exclusively in the classroom or lecture hall, but most include a lab section or, at least, some experimental or observational projects. If your course includes lab, experimentation, or even lecture demonstration, you can be certain that the work you do in these sections will show up on exams. (In some introductory physics sequences, lab sections have their own exams.) Take notes, and be certain that you understand not just the process of the experiment or observational exercise, but the significance of it as well.

Be There

Physics is a *basic* science, which looks at ordinary things in what may strike you as extraordinary ways. It is an *orderly discipline*, resting on clearly stated assumptions with clearly stated objectives and goals.

The *orderly* quality of physics implies the necessity of proceeding through the subject step-by-step, making certain that you understand each step before you move on to the next. Learn the vocabulary *well*. Understand the basic principles, theories, and laws *well*. Learn to work with graphs, diagrams, equations, and drawings *well*. Be certain that you understand all experimental or observational work fully, in process as well as significance.

- ◆ Don't skip over textbook material.
- ◆ Don't just start reading words without taking the time and making the effort to understand them.
- ◆ Don't just glance at the graphs, diagrams, equations, and other illustrations you encounter. Interpret them by translating what you see into words.

Be there. When you study your textbook, *study* it. Practice the concepts you read about. Physics is as much a set of observational skills as it is an intellectual pursuit. Acquiring and developing any skill requires active practice, not just passive reading.

Be there—in class. Don't skip classes. Attend lectures. Take notes. Ask questions. Most instructors believe that *most learning takes place in the classroom*. This means actively listening to lectures, actively participating in class discussions, and—very important—asking questions *as they occur to you*. If you don't understand something in a lecture, ask about it as soon as possible. Don't let material pass you by! Doing so will make it that much harder to take the next step.

More Than a Set of Problems

We've all known people who are so intensely detail oriented that, in pursuit of every last fragment of minutiae, they consistently miss the big picture. As the old saying goes, they can't see the forest for the trees. Physics, which requires learning many concepts, new terms, and sets of mathematical expressions, unfortunately presents this very trap. You can avoid falling into it by trying to think of physics as something more than the sum of a series of concepts, equations, words, and diagrams. These are important, but they alone are not the purpose of physics. You are getting a peek at a field that sets as its goal nothing less than understanding how the physical universe works. Moreover, it is a discipline that has been at the forefront of modern science, not only in theoretical fields, but in such applied fields as atomic energy, electronics, and space exploration.

Try to keep the truly breathtaking scope of physics in mind as you wrangle with a stubborn equation. It will drive and invigorate your work, and it will help you to bring together all those many constituent parts—formulas, vocabulary, tables, graphs, diagrams, drawings, equations—into a much more meaningful picture.

PART TWO

STUDY GUIDE

CHAPTER 5

INTRODUCTION TO PHYSICS: THE MAJOR TOPICS

I N THE PRECEDING CHAPTER, WE BROADLY DEFINED PHYSICS AS DEALING WITH THE structure of matter and with the interactions between the fundamental constituents of the observable universe. The ultimate goal of physics is to formulate a comprehensive set of principles that bring together and explain all the apparently disparate and diverse phenomena that surround us. We also mentioned that physics describes the physical changes of objects in terms of the dimensions of length, time, mass, velocity, acceleration, electric current, temperature, amount (of a substance), and luminous intensity. All of this forms the foundation of physics. Beyond this, the field may be divided into two very broad categories: *classical physics* and *modern physics*.

Classical Physics: Science of the Everyday

Much of what you will learn in most introductory physics courses has been studied and known since before 1900. Classical physics encompasses *classical mechanics*: the study of *waves and sound, thermodynamics*, the study of *electricity and magnetism*, and the study of *light*.

Classical Mechanics

The terms *classical mechanics* and *Newtonian mechanics* are synonymous, because the landmark formulations and principles of the motion of material objects were established by the British physicist Sir Isaac Newton (1642–1727), one of the most remarkable minds the world has ever known.

Newtonian mechanics encompasses:

♦ **Kinematics.** The analysis of the positions and motions of objects as a function of time, without regard to the causes of motion. At its most basic, kinematics is studied in one dimension (motion forward or backward along a number line). More complex is kinematics in two dimensions, which takes into account trajectories.

♦ **Dynamics.** Newton ventured beyond kinematics to study motion in relation to its cause—that is, *force.*

♦ **Work and energy.** These twin concepts are closely related to dynamics. *Work* is what happens to an object when a force is applied to it, moving it through distance. *Energy* is the ability to do work.

♦ **Rotational motion.** Ascending further in complexity is the study of rotational motion. This study encompasses the principles of *planetary motion*, first formulated by the German astronomer Johannes Kepler (1571–1630), and which also form the basis for Newton's law of universal gravitation.

♦ **Elasticity and harmonics.** The modes of motion listed so far all concern a theoretical idealization called a *rigid body*—a body that remains absolutely rigid (undeformed) regardless of the amount of force applied to it. The physics of motion moves closer to the real world in the study of elasticity, the relationship between solid-body deformations and the forces that cause them. Related to elasticity is *harmonic motion*, the motion of vibrating bodies.

♦ **Fluids.** Classical mechanics also studies the behavior of fluids in motion—a fluid being defined as a substance that takes the shape of its container (thus, for a physicist, a gas may be treated as a fluid).

Waves and Sound

As understood in classical physics, the subjects of waves and sound are very closely related. Classical wave theory (as we will see in Chapters 10 and 11) proved inadequate to explain the behavior of light.

Wave theory is chiefly concerned with describing the motion of waves. Sound is studied as a manifestation of wave motion.

Thermodynamics

Although the term *thermodynamics* suggests the study of heat, this branch of classical physics goes well beyond heat (as common sense understands it) to explain some essential properties of matter, especially the relationship between the behavior of matter on the macroscopic level and on the atomic level.

Electricity and Magnetism

At its simplest, physics studies electricity as *electrostatics*—the familiar phenomenon associated with static electricity, stationary electric charges, and electric fields. Beyond this, classical physics also encompasses:

◆ The properties and behavior of *electrical circuits*

◆ *Electromagnetic forces and fields*

◆ *Electromagnetic induction*

Light and the Electromagnetic Spectrum

The nature and behavior of light has long been of great interest. Newton supposed that light is essentially a stream of particles, whereas Christian Huygens (1629–1695) proposed a wave theory. Neither explanation alone proved adequate, and, ultimately, a theory combining notions of light as behaving like a stream of particles *and* a wave marked the transition from classical to modern physics.

Classical physics describes the behavior of light in terms of geometry (*geometrical optics*) and wave theory (*wave optics*). Both are necessary to account for the behavior of light in its entirety.

The nineteenth-century British physicist, James Clerk Maxwell (1831–1879), demonstrated that visible light occupies merely one (rather narrow) range of frequencies on a continuous *electromagnetic spectrum*, which encompasses energy from gamma rays (extremely short wavelengths), to x-rays (longer), to ultraviolet (longer still), to visible light (from violet at the short-wave end of the visible spectrum to red at the long-wave end), to infrared (longer wavelengths than visible light), to radio waves.

Modern Physics: Reality from a New Perspective

The subject of light forms an apt transition from classical physics to modern physics. It is not that classical physics is wrong and modern physics is right, but that modern physics looks at reality from a new frame of reference, a perspective beyond the everyday, essentially macroscopic world. Modern physics studies nature on the level of the behavior of atomic and subatomic particles (the behavior of which is not accounted for in Newtonian physics). It also studies what happens at very high velocities, approaching the speed of light, and how reality is structured on the very largest—interstellar—scales.

Introductory physics courses generally approach modern physics in three broad areas: *quantum mechanics, nuclear physics*, and *relativity*.

Quantum Physics

By the end of the nineteenth century, the inadequacy of classical physics to explain certain phenomena (such as the behavior of matter at the atomic and subatomic levels and motion near the speed of light) had become apparent. Quantum physics (Chapter 13) grew out of the work of Max Planck (1858–1947) to address the short-comings of classical physics. Quantum physics holds that energy is always exchanged in units of fixed size (quanta). Conceptualizing energy transmission in terms of such "packets" revolutionized physics by enabling it to account for the behavior of matter at the atomic and subatomic levels.

Nuclear Physics

Like quantum physics, nuclear physics deals with phenomena at the atomic and subatomic levels, but focuses on energy exchanges on this level: between atomic nuclei and electromagnetic fields, between tightly bound electrons in heavy atoms and electromagnetic fields, and between fragments of nuclei. A dramatic by-product of nuclear physics was the liberation of the tremendous energies that bind the particles of atoms together. One result of this research was fission and fusion weapons.

Relativity

Chapter 15 is devoted to the most profound and revolutionary theory of modern physics, relativity, first presented by Albert Einstein (1879–1955) in 1905. The 1905 theory, called special relativity, explained physical reality at or near the speed of light. General relativity, developed later, is a theory of gravitation that greatly extends Newton's understanding of the phenomenon. Einstein's concept of general relativity holds that massive matter does not merely attract other massive matter (as Newton said), but that all matter warps the space around it. Even photons of light feel the effect of gravity, actually bending with the contour of space itself. Relativity theory unites such *apparently* diverse phenomena as gravity, time, light, and space.

Road Map

Remember, this book is a guide to help you use midterms and finals, inevitable facts of college life, to focus study of your physics course. It is not a comprehensive intro-duction to physics, and it is certainly not a substitute for reading your textbooks, attending lectures, and actively participating in class discussion. This chapter and the others in this part of the book should serve to point out the most prominent physics landmarks you will encounter on your trip through the course. We have already inventoried a few of physics' most basic assumptions and areas of inquiry.

Here are the major themes and topics that are built on these assumptions and areas and that are reflected in the next 10 chapters:

Motion and Newton's Laws of Motion
• Kinematics in One Dimension
• Distance and Position
• Speed and Velocity
• Equations
• Speed
• Acceleration
• Kinematics in Two Dimensions
• Dynamics: Newton's Laws of Motion
• Newton's First Law
• Newton's Second Law
• Newton's Third Law

Principles of Momentum, Work, Energy, and Conservation
• Conservation of Momentum
• Work, Power, and Energy
• Conservation of Energy

Rotational and Planetary Motion
• Rotational Inertia
• Angular Velocity and Angular Acceleration
• Torque
• Angular Momentum
• Planetary Motion
• Newton and Universal Gravitation

Thermodynamics
• The Nature of Heat
• Another Measure of Heat
• Heat Flow
• Thermal Expansion
• Heat and States of Matter
• Evaporation and Condensation
• Boiling
• Melting and Freezing
• Sublimation

- Behavior of Gases
- The Laws of Thermodynamics
- Implications of the Second Law of Thermodynamics
- Entropy

Waves and Sound
- What Is a Wave?
- Anatomy of a Wave
- Wave Interaction
- Doppler Shifts
- Sound
- Pitch
- Intensity
- Force Vibrations
- Resonance
- Beats

Electricity, Magnetism, and Electromagnetism
- Electrostatics
- Electrical Charge
- Electrical Forces
- Transmission of Charges
- Producing an Electrostatic Charge
- Electric Fields
- Electrical Potential Energy
- Electrical Potential
- Electric Current and Electrical Circuits
- Current Unit
- Current Flow
- Circuits
- Magnetism
- Poles
- Fields
- Electromagnetism
- Electromagnetic Induction

Light and the Electromagnetic Spectrum
- What Is Light?

- The Electromagnetic Spectrum
- What Makes Color?
- Speed of Light
- Orientation of Light Waves
- Geometrical Optics
- Reflection
- Refraction
- Wave Optics

Quantum Physics
- Quantum Mechanics
- Black-Body Radiation
- Photoelectric Effect
- Particle-Wave Duality and Its Implications
- The Bohr Atom
- De Broglie Waves
- Heisenberg's Uncertainty Principle

Nuclear Physics
- Radioactivity
- Nuclear Structure
- Fission
- Fusion
- Chain Reaction
- More Particles

Relativity Theory
- Maxwell's Electromagnetic Equations
- Michelson-Morley Experiment
- Special Theory of Relativity
- Simultaneity
- Time Dilation
- Mass Increase
- Mass-Energy Equivalence
- General Relativity
- Space-Time
- Black Holes
- Cosmology

CHAPTER 6

MOTION AND NEWTON'S LAWS OF MOTION

Y OUR INTRODUCTORY PHYSICS COURSE MAY WELL BEGIN WHERE YOU BEGAN as a toddler, pushing objects around and watching what happens when they move. The difference between toddlerhood and physics is that, in physics, motion is described exactly and mathematically to predict effects.

Kinematics in One Dimension

Kinematics analyzes the motions and positions of objects as a function of time, but without reference to the cause of the motion.

Distance and Position

◆ **Distance.** Expressing in units the linear space between two points is *distance*. Distance is a *scalar* quantity, because it carries information about magnitude (number of units of linear space), not direction. *Objects A and B are 2 meters apart* is a scalar expression of distance.

◆ **Position.** Combining distance with position is the expression of a *vector* quantity. Vectors always refer to some origin. *Object A is 2 meters east of Object B* is a vector expression.

Speed and Velocity

◆ **Speed.** Like distance, speed is scalar. Speed is the rate of change of the distance from a starting point over time.

◆ **Velocity.** Like position, velocity is a vector. It is speed combined with direction. It is the rate of change of position.

Acceleration

◆ **Acceleration.** The rate of change of velocity is acceleration.

Acceleration is a vector and always has reference to direction. Stepping on your car's accelerator provides an example of acceleration in the direction of velocity, whereas stepping on the brake is an example of acceleration in the opposite direction. A motorist would call the first instance *acceleration* and the second instance *braking*, but, to a physicist, both are examples of acceleration—just in different directions.

Likewise, steering, for a physicist, is also acceleration. The direction in this case is more or less perpendicular to the velocity of the object. Such perpendicular acceleration produces a change in direction, not speed.

Equations

All of these relationships—as, indeed, all relationships in physics—can be most usefully expressed mathematically as equations. Because the chapters in this part of *Ace Your Midterms and Finals: Introduction to Physics* are intended only as a brief overview of the kind of material you are most likely to cover in an introductory physics course, we will look at far fewer equations that you will find in your textbook. Principally, we'll be surveying concepts, not calculations. However, simple motion does provide opportunity for introducing equations and how they relate to physical reality.

Speed

Remember, speed (v) is the rate of change of an object's distance (d) from a starting point as time (t) passes:

$$v = \frac{d_2 - d_1}{t_2 - t_1}$$

Note the subscripts: d_1 is the distance corresponding to t_1, and d_2 is the distance corresponding to t_2.

Acceleration

Acceleration (a) is the rate of change of velocity (v):

$$a = \frac{v_2 - v_1}{t_2 - t_1}$$

Kinematics in Two Dimensions

You will doubtless spend some time studying motion in one dimension, becoming accustomed to working with equations involving speed, velocity, and acceleration. Early in the course, you may also study *free fall* situations, the effects of acceleration produced by gravity. However, our world is three-dimensional, and the next step up in complexity, more closely approximating the real world, is the exploration of kinematics in two dimensions.

An example of a real-world motion is a ball tossed into the air. Its motion can be viewed as the combination (*superposition*) of two motions for a two-dimensional view: a vertical motion and a horizontal motion (the ball travels up and down, and also covers a certain horizontal distance). Superposing these two dimensions of motion results in a description of the object's *trajectory*, its path.

Mathematically describing motions in two dimensions is more complex than describing simple one-dimensional motions. The equations involved, however, do not go beyond elementary vector algebra, and two-dimensional kinematics problems are a very good way to sharpen up (or dust off) the math skills necessary for work later in the course.

Dynamics: Newton's Laws of Motion

Kinematics is confined to the study of the variables involved in motion, but without reference to the cause of motion, *force*. Sir Isaac Newton formulated three laws of motion, which, together with algebra, may be used to analyze the relationships of acceleration and force, as well as *inertia*. The analysis of motion encompassing the relationships between these variables is called *dynamics*.

Newton's First Law

Newton's first law of motion states that the velocity of an *isolated* object does not change as time passes. Newton called this property of objects *inertia*.

◆ An *isolated* object is free from outside forces, including friction. On earth, this is an idealized condition, but a spaceship, its engines off, coasting between the earth and the moon in the vacuum of space is very close to a real-world example of an isolated object.

◆ Note that inertia relates to the tendency not to change. Thus, a motionless isolated object tends to remain motionless, and an isolated object in motion tends to remain in motion at an unchanging velocity.

Inertia is measurable in terms of *mass*, which is, in effect, a measure of the amount of matter in an object. By convention, mass is measured in kilograms (kg).

Newton's Second Law

If a net *force* acts on an object, it will cause acceleration of that object. This may be expressed by a simple equation: $F = ma$, where F is force, m is mass, and a is acceleration. Physicists may use a Standard International (SI) unit to measure force, the newton (N), or the common English unit, the pound (lb). A newton is the equivalent of about 0.22 lb. One newton is the magnitude of force required to accelerate a 1-kg object at a rate of 1 meter per second squared (1 m/s^2).

Newton's Third Law

For every action, there is an equal and opposite reaction; that is, if object A exerts a force on object B, object B must at the same moment exert an equal and opposite force on object A. This is a remarkable insight. The exertion of force (in the form of gravitation) is a property of all matter that has mass. If you jump down from a height, the force of gravity pulls you to earth, right? Actually, the mass of the earth creates the gravity that attracts your mass, but your mass also exerts a force on the earth. Both you and the earth accelerate as a result of these forces; however, the earth is so much more massive (contains so much more matter) than you are, that your effect on the earth, although real, is imperceptible, whereas the earth's effect on you is quite apparent.

Newton's laws are understood through equations and also through *force diagrams*, also called *free-body diagrams*, examples of which you will see in most of the sample exams in this book. Free-body diagrams help you visualize the forces acting on a given object. Often, a force diagram is a helpful (or even necessary) first step before writing an equation to express the relationship between forces and acceleration.

Note that mass and weight are two different physical quantities. Mass is a measure of inertia, the amount of matter in an object. Mass does not vary with gravitational field. In contrast, weight is an expression of force and is a function of gravity: Weight (W) = mass (m) times the acceleration due to gravity (g)—$W = mg$. Thus the mass of a given object is the same, whether it is on the earth or the moon, whereas the weight of that object will vary with the force of gravity (and will weigh on the moon one-sixth of what it weighs on the earth).

CHAPTER 7

PRINCIPLES OF MOMENTUM, WORK, ENERGY, AND CONSERVATION

FROM AN OVERVIEW OF NEWTON'S THREE LAWS OF MOTION, YOU WILL LIKELY PROceed to a closer investigation of the second law, dealing with force. A mass that is acted on by a net average force (F) will undergo an average acceleration expressed by $F = ma$. The product of F multiplied by the time of contact is called the *impulse*:

$$F(\Delta t) = mv_f - mv_i$$

where v_i is the initial velocity and v_f the velocity after the force is no longer in contact with the body. Impulse is measured in newton-seconds (N–s).

The right side of the impulse equation is the *linear momentum* of the object. Linear momentum is the product of mass times velocity:

$$p = mv$$

where p is linear momentum. Linear momentum is measured in kilogram meters/second (kg m/s). Momentum, in effect, is a measure of how much force is required to stop a moving mass or to put it into motion.

Conservation of Momentum

The *law of conservation of momentum* states that, in the absence of an external force acting on a system, the total (*vector sum*) momentum of the system remains constant. This makes it possible to analyze interactions between systems of objects because before, during, and after a collision between two (or more) isolated objects, the sum of their momenta is conserved, is constant. Using the law of conservation of momentum, you will examine what happens when objects collide in one or two dimensions.

Work and Energy

An impulse produces a change in momentum, whereas work produces a change in energy. *Work* is done on an object when an applied force moves it through a distance. *Energy* is the ability to do work. *Power* is the rate at which work is done. It is important to recognize the differences among these related phenomena.

Work

Work is expressed mathematically as

$$W = Fx \cos \theta$$

where F is applied force, x the distance moved (also called *displacement*), and θ the angle between the direction of the force and the displacement. In contrast to momentum and impulse, which are vector quantities, work is scalar. The Standard International (SI) unit for work is the joule (J), which is a newton-meter (kg m/s²).

Power

As the rate at which work is done, *power* (P) is expressed mathematically as

$$P = \frac{W}{t}$$

where P is power, W work, and t the time interval during which the work is done. The SI unit of power is the watt (W). One watt is the equivalent of 1 J of work done in 1 s (second):

$$1 \text{ J/s} = 1 \text{ W}$$

Energy

Energy, the ability to do work, may be *kinetic* or *potential*. Kinetic energy is the energy of an object in motion. It is readily computed by calculating the work (W) required to accelerate an object at a constant rate (a) from rest to speed v in a particular time duration (t) during which the object moves through a distance (d). Kinetic energy is dependent on the mass and speed of the object and is independent of how the object attained that speed.

Potential energy is stored energy: the ability of a system to do work due to its position or internal properties. For example, a coiled spring possesses potential energy, which becomes kinetic energy when the spring is released. (The type of potential energy stored in a spring is called *elastic potential energy*.) A ball at the top of an incline likewise possesses potential energy, which is released as the ball rolls down the incline. A storage battery also possesses kinetic energy due to its chemical properties; the energy becomes kinetic when an electrical circuit is closed and a chemical reaction within the battery produces electrical current.

Conservation of Energy

One of the most important (and most familiar) principles of physics, the *law of conservation of energy* states that energy is neither created nor destroyed, but can only be transformed from one form to another in any given isolated system. As with conservation of momentum, conservation of energy allows straightforward analysis of a physical system in which energy is transformed.

In the real world, on earth, systems in which *mechanical energy* is conserved don't exist. That is, outside forces always intrude, typically in the form of *friction*. Nevertheless, it is not necessary to abandon the concept of conservation of energy even in this case. Friction transforms some of the mechanical energy into *heat*, another *form* of energy. Thus, it could be demonstrated that the sum of all forms of energy (mechanical + heat) is conserved, even in the real world.

Note that conservation of energy and the transformation of energy are of great practical concern. Some real-world mechanical systems manage to achieve a high degree of conservation of mechanical energy by reducing friction to a minimum with lubricants and careful craftsmanship. By the eighteenth century, for example, clockmakers were constructing pendulum clock movements with such skill that some clocks could run accurately, on their own, for days before a significant amount of mechanical energy had been transformed by friction into heat.

In some theoretical situations, for a physicist, energy is energy, regardless of the form it takes. But to an engineer, for example, the form energy takes is of great importance. *Efficiency* may be defined as the useful work obtained from a machine divided by the energy put into it. No real-world machine is 100 percent efficient. Think of even the most expensive German automobile engine. The *useful work* it produces is motion. But it also produces a great deal of work that is not useful, in the form of heat and noise. It hardly approaches a theoretically high level of efficiency.

The term *elastic* bears some further consideration. In most introductory physics courses, you will work with situations involving the collision of two or more objects. The idealized situation is an *elastic* collision, in which colliding objects interact without losses (transformation) of kinetic energy due to friction or to the deformation of an object. (Auto manufacturers talk about "crumple zones" as structural safety features that absorb the force of a collision by crumpling, or *deforming*.) In the real world (at least at the macroscopic level), most collisions are *inelastic*, with some energy lost (transformed) as a result of friction and/or deformation. Mathematically describing inelastic collisions is much more complex than doing the same for elastic collisions.

CHAPTER 8

ROTATIONAL AND PLANETARY MOTION

ROTATIONAL MOTION IS THE NEXT STEP UP IN COMPLEXITY FROM LINEAR MOTION, and the concepts associated with rotational motion are built upon those that apply to linear motion. Throughout your experience with introductory physics, you will find that it is critical to understand one level of a concept or phenomenon to progress to the next level. More than many other subjects, physics builds upon foundations. To understand rotational motion, you must understand the principles of linear motion.

Rotational Inertia

As a body in linear motion resists change in velocity, so a body in rotational motion resists change in its rate of rotation as well as in the orientation of its axis of rotation. The quantitative measure of this *rotational inertia* is the *moment of inertia*. The basic equation is

$$I = mr^2$$

where *I* is the moment of inertia, *m* is the mass of the object, and *r* is the distance from an axis about which the object moves in a circle.

This equation is useful only for an object of very small mass. In more massive objects (*extended objects*), calculating *I* requires computing the *average value* of the square of the distance from the mass to the rotation axis. How the mass of an object is *distributed* has an effect on *I*. For example, you might observe the difference in speed with which two cylinders roll down an incline. The two cylinders, say, are identical in shape, size, and mass, but one is a light solid and the other a cylindrical ring made of heavy material. Despite the properties they share, the mass in these two bodies is distributed differently; therefore, the ring, which carries all of its mass far from its center, resists an increase in speed more effectively than the solid cylinder. The moment of inertia (I) of the ring is greater than that of the solid cylinder.

Angular Velocity and Angular Acceleration

The velocity of rotational motion is measured in terms of *angular velocity*. The *angular displacement* of a rotating wheel is the angle between the radius at the beginning and the end of a given interval of time as expressed in SI units called *radians*. The average *angular velocity* (ω) is measured in radians per second and is expressed thus:

$$\omega = \frac{\text{angular displacement}}{\text{elapsed time}} = \frac{\theta}{t}$$

Angular acceleration (α) is expressed in this way:

$$\alpha = \frac{\text{change in angular velocity}}{\text{elapsed time}} = \frac{\omega_f - \omega_0}{t}$$

It is measured in radians per second per second: rad/s^2.

Torque

Torque is the product of the part of a force F applied to an object that is perpendicular to its rotation axis, F', times the perpendicular distance ($d_{\text{perpendicular}}$) from the line along which the force acts to the rotation axis about which the torque is computed:

$$\tau = F' d_{\text{perpendicular}}$$

It is far easier to conceptualize torque intuitively. Push a door close to its hinges, and you will find that it is difficult to open the door. Push farther from its hinges—where the knob or handle is—and you will have little trouble.

The action of torque on rotation produces rotational analogs to Newton's three laws of linear motion:

1. The angular acceleration of an object not acted on by any net torque is zero.

2. The angular acceleration of an object is the net torque acting on it divided by its moment of inertia.

3. If object A exerts torque on object B, B exerts an equal and opposite torque on A.

Angular Momentum

The rotational analog of linear momentum is *angular momentum* (*L*). Angular momentum is the product of the moment of inertia times the rate of rotation of an object:

$$L = I\omega$$

◆ The angular momentum of an object not acted upon by a net torque is conserved.

◆ In the presence of net torque, the rate of change of angular momentum is equal to the net torque.

◆ A torque parallel to an angular momentum that points along the rotation axis increases or decreases the angular momentum without changing direction.

Planetary Motion

Directly flowing from the basic concepts of rotational motion are the principles that explain the grandest rotational motion of all, that of the planets.

The Danish astronomer Tycho Brahe (1546–1601) was a brilliant observational astronomer who amassed a mountain of data recording the movement of the planets. Late in his life, he took on a student, the German astronomer Johannes Kepler (1571–1630). When Tycho died, Kepler dug into the data Tycho had left. After years of study, Kepler reduced Tycho's mountain to three simple laws:

1. The planets orbit the sun not in perfect circles, but in elliptical orbits, a flattened circle drawn around two *foci* instead of a single center point.

2. An imaginary line connecting the sun to any planet would sweep out equal areas of the ellipse in equal intervals of time. This explains the observed (and, until Kepler's time, bewildering) variation in speed with which planets travel. They move faster when they are closer to the sun.

3. The square of a planet's *orbital period* (the time needed to complete one orbit around the sun) is proportional to the cube of its *semimajor axis* (in effect, the cube of the average radius of its orbit).

With Kepler's three laws, it is possible to build a scale model of the solar system, illustrating the correct elliptical shapes of the orbits (those of all but Mercury and Pluto are *almost* circular) and the sizes of the orbits, expressed not in *absolute* miles, but relative to a portion of earth's orbit. (Because Kepler did not know the absolute size of earth's orbit, he could not express the other orbits in terms of absolute distance.)

Newton and Universal Gravitation

Newton thought long and hard about Kepler's picture of the solar system and con-cluded that the orbits of the planets were subject to the same gravitational force that draws a falling apple to earth. The force of gravity exerted by objects upon one another weakens as the distance between those objects increases. Many forces, such as gravity, decrease in proportion to the square of the distance. This *inverse-square law* means that the force of gravity mutually exerted by two objects, say, 10 units of distance apart is 100 times (10^2) weaker than that exerted by objects only 1 unit apart—yet this force never reaches zero.

Proceeding from the inverse-square law, Newton concluded that the gravitational pull between two objects is directly proportional to the product of their masses (mass of object A times mass of object B) and inversely proportional to the square of the distance separating them. This, in a nutshell, is Newton's law of gravity.

Consider the sun and any one planet. These two bodies exert gravitational pull on one another, though the sun's mass is much greater than the planet's and, there-fore, the planet's acceleration is much greater than that of the sun. Newton's first law of motion dictates that the planet will keep moving at a constant speed and in a straight line *unless* some external force acts on it. That force is the mutual attraction of the sun and the planet, a force that, in effect, competes with the planet's inertia, its tendency to keep going at a constant velocity and in a straight line.

If the planet lacked sufficient velocity in its orbit, the gravitational pull would "win," and the planet would be drawn closer and closer to the sun. It would *fall* into the sun. If, however, the planet had sufficiently great velocity, it might overcome the sun's gravitation (reach *escape speed*) and more closely approach a straight-line course out into space. It would fly out of orbit. But at a certain velocity, the planet neither falls into the sun nor shoots off into space. Instead, its path is bent in con-tinuous orbit around the sun. The orbit may not be perfectly circular, because the gravitational forces of other planets and objects come into play, competing with the dominant pull of the sun, but it is an orbit nonetheless. And so the planets of the solar system move, without falling into the sun or flying off into space.

Consider the relatively simple case of the moon orbiting the earth. The orbit, though elliptical, is very nearly circular. In effect, the moon falls *around* the earth, having an acceleration just sufficient to keep it in orbit—neither falling into the earth nor flying off into space. This is not a lucky chance, but proof that the strength of a gravitational field varies with the distance r from the source mass: I/r^2.

CHAPTER 9

THERMODYNAMICS

THERMODYNAMICS, AS THE TERM SUGGESTS, DEALS WITH PHENOMENA OF TEMPERA-
ture, heat, and heat transfer, but this description is deceptive. In fact, thermodynamics explains much about the properties of matter on the
macroscopic level and draws correlations between properties on this level
and properties on the atomic and molecular levels.

The Nature of Heat

At first glance, the subject of thermodynamics may seem a great leap from the subject of classical mechanics, which we have just briefly surveyed. Actually, thermodynamics is based on two results from mechanics:

1. Mechanical energy is apparently lost in motion with frictional forces; however,
   heat is considered a form of energy, so that it can be seen that the mechanical
   energy "lost" is really *transformed* into heat.

2. Within matter, atoms and molecules may vibrate or fly about—mechanical
   kinetic energy that produces heat. The random motion of atoms and molecules in matter makes matter seem hot.

Beginning students of physics must take care when thinking about heat. We intuitively think of it in terms of our senses—what *feels* hot or cold. As physicists see it,
heat is a property of all matter, because the motion of atoms and molecules produces heat. Only at absolute zero (an ideal state) is there an absence of heat, because

there is zero atomic motion.

Temperature is a quantitative measure of the tendency of *heat flow*, which occurs when *thermal energy* leaves an object and flows into another object. Temperature is *not* a direct measure of thermal energy, because thermal energy is a function of volume; that is, 1000 cubic centimeters (1 liter) of boiling water has 1000 times more thermal energy than 1 cm³ of boiling water, even though both are at a temperature of 100° C. Indeed, 1 L of ice-cold water has more thermal energy than 1 cm³ of boiling water. Importantly, heat flow is always from the hotter object to the colder (if they are in thermal contact), regardless of the relative volumes of the two objects. This is true even if the thermal energy of the cooler substance is greater than that of the hotter. Thus, the heat from the 1 cm³ of boiling water would flow into the liter of ice water, if the two were mixed, even though the liter of cold water has greater thermal energy.

Another Measure of Heat

Temperature may be measured on various scales, the most common being the Fahrenheit scale, the Celsius scale, and the Kelvin scale. But these scales do not measure actual *units* of heat. By convention, heat units are the amounts of heat required to produce standard increases in the temperatures of standard substances. One *calorie* (1 cal) is the heat required to increase the temperature of 1 g of water 1 Celsius degree.

> Note that *Calorie*—with a capital C—denotes 1 kcal, 1000 calories (lowercase c), and is the unit conventionally used to measure the energy content of foods.

Heat units are necessary because it takes more heat to raise the temperature of, say, 1 g of water than it does to raise the temperature of the same quantity of iron. Heat units provide *specific heats*, which describe the differences between materials.

How much energy is represented by a *calorie*? To determine the mechanical equivalent of heat, it is necessary to measure the work done in a situation in which all of the work is converted to thermal energy. The physicist James Joule (1818–1889) did just this and arrived at a figure of 4.19 J as the energy represented by a single calorie.

Heat Flow

Newton formulated a *law of cooling*, which states that heat flows from a hot body to a colder one proportionate to the difference in temperature between the two bodies. This is why a hot liquid, such as coffee, falls in temperature rather rapidly at first, then more slowly as it cools. When it is piping hot, the difference in temperature between it and the room is much greater than when it has cooled. The rate of heat flow is reduced proportionately.

Heat flows as a result of:

♦ **Contact.** When two different materials are in direct contact, the moving surface molecules of one material collide with the surface molecules of the other material, thereby transferring kinetic energy.

♦ **Conduction.** Some materials conduct heat more efficiently than others. Metals, for example, are good conductors and, therefore, poor insulators. Such substances as wood and cloth are inefficient conductors and, therefore, efficient insulators.

♦ **Convection.** The expansion of liquids and gases with heating produces *convection currents* by which warm material of lower density rises while colder, denser material sinks. The result is heat transfer without physical contact.

♦ **Radiation.** Convection currents are not the only means of transferring heat in the absence of physical contact. Even in a vacuum (where convection is an impossibility), heat is transferred: hot matter loses heat to cooler matter. The transfer is in the form of electromagnetic radiation.

Thermal Expansion

For the most part, matter expands when its temperature rises. The increase in kinetic energy means that atoms collide more frequently and push each other harder, thereby occupying more space. The matter expands.

An exception to this case is water, which does expand as its temperature rises above 4°C, but also expands when its temperature *decreases* from 4°C to freezing, 0°C. This anomaly is a consequence of how water molecules bond with one another.

Heat and States of Matter

In dense, solid matter, a rise in temperature produces thermal expansion. In a liquid, such a rise may produce thermal expansion as well as *evaporation*, which is the result of molecules near the surface of the liquid moving with such kinetic energy that they escape the attractive force of the molecules deeper below the surface.

Evaporation and Condensation

Molecules with the greatest kinetic energy leave the liquid (evaporate), while those with less kinetic energy remain behind. Thus *evaporation* is a cooling process, the highly energetic molecules carrying heat away from the liquid. This is what happens when we sweat. Water is secreted through the pores, some of which evaporates, carrying away body heat.

The complement to evaporation is *condensation*. When gas molecules collide with a cool surface, they may lose so much energy that they cannot move away from the surface. As more molecules accumulate and "stick" to the surface, a layer of liquid forms.

Boiling

Boiling results when sufficient heat is applied to a liquid to cause a large number of molecules to become sufficiently energetic to escape from the liquid state, even though they are not near the surface. The result is bubbles of gas within the liquid, which rise. Boiling actually cools the liquid that is left behind; however, you wouldn't immerse your hand into a pot of boiling water, because, to keep the water boiling, we usually replace any heat that is lost by maintaining a flame under the pot.

Melting and Freezing

Molecules vibrate in all matter. If they stay relatively close to their average positions, however, the material remains, to outward appearances, solid and inert. Heat the material sufficiently, and enough kinetic energy is transferred to the molecules to move them past one another with ease. Individual molecules thus move relatively far from their original "frozen" positions, and the solid substance melts. Remove the heat, and the molecules begin to lose kinetic energy in their collisions, they slow down, and the substance freezes again.

Sublimation

Generally speaking, a given material may exist as a solid until sufficiently heated to melt it into a liquid. If the liquid is heated further, it may evaporate or boil into a gas. Under certain circumstances, molecules in a solid may gain sufficient energy to be transformed directly into a gas, without passing through the liquid state. This is called *sublimation*.

Behavior of Gases

The relationship among pressure, volume, temperature, and quantity in gases is of interest to physicists because this relationship clearly links the behavior of matter at the subatomic level to the observed behavior at the macroscopic level.

The behavior of real gases may be modeled through an idealization, an *ideal gas*, which consists of identical particles that interact only occasionally as if they were elastic (nondeforming) billiard balls. Two important laws apply to the behavior of an ideal gas:

◆ **Boyle's law.** If a gas is compressed while the temperature remains constant, the pressure varies inversely with the volume. The product of the pressure and volume is a constant.

◆ **Charles/Gay-Lussac law.** For a constant pressure, the volume of a gas is directly proportional to the Kelvin temperature.

By combining these laws, the *ideal gas law* is derived, forming the mathematical basis of a *kinetic theory of gases*, which mathematically expresses the relationship between temperature and the average kinetic energy of gas molecules. Pressure is seen as proportional to the number of molecules per unit volume and to the average linear kinetic energy of the molecules. Temperature, therefore, is a direct measure of the average molecular kinetic energy of an ideal gas.

The Laws of Thermodynamics

Thermal energy is the energy of the motion of molecules in matter. Heat flow is the transfer of thermal energy from one object to another. Work can produce thermal energy, and thermal energy can produce work. The *first law of thermodynamics* expresses this relationship mathematically: The heat (Q) added to a system divides (in some way) into an increase in the internal energy of the system (U) and work (W) done by the system. The equation is

$$Q = \Delta U + W$$

The *second law of thermodynamics* is a statement of heat flow: heat always flows from a hotter object to a colder one.

Implications of the Second Law of Thermodynamics

The Scottish inventor James Watt (1736–1819) devised a steam engine that converted heat to mechanical energy with sufficient efficiency to be of great practical value. Among those who studied Watt's engine was the French scientist Nicolas Leonard Sadi Carnot (1796–1832), who asked (and sought to answer) just how efficient a heat engine could be. That is, how much useful work could a heat engine produce per unit of heat energy put into it?

Carnot posited an engine so efficient that it violated the second law; that is, it caused the flow of heat from cold to hot without any external energy input. If this is held to be impossible, then it must be true that *any* engine that takes in heat from a hot source or place *must* transfer some of that heat to a colder place. It cannot convert all of its heat input into mechanical work. As Carnot defined it, efficiency (ε) is the quantity of work output divided by the quantity of heat input. Expressed in terms of the second law of thermodynamics, the greatest possible efficiency for a heat engine is:

$$\varepsilon_{ideal} = \frac{T_{hot} - T_{cold}}{T_{hot}}$$

Note that the temperature (T) must be absolute, expressed in degrees Kelvin. *Absolute temperature* takes 0° as the temperature at which molecular motion ceases—that is, the point of 0 energy. Other scales, such as the Celsius or Fahrenheit, are indexed to arbitrary phenomena, such as the temperature at which water freezes or at which it boils.

Carnot demonstrated a practical limit to the efficiency of a heat engine. The implication is profound on an immediately practical level (part of what you pay for in gasoline is "wasted" in the production of heat and noise rather than the production of useful motion), but even more profound for what it says about the nature of the universe. It is impossible to transform all of the *random* kinetic energy of molecules in matter into useful *orderly* work. Randomness—disorder—is an integral part of nature and cannot be undone by human intervention.

Entropy

Nor is it that randomness is simply a part of reality. A further implication of the second law of thermodynamics is that isolated systems tend toward a greater degree of disorder. The physics field called *statistical mechanics* quantitatively measures disorder as *entropy*. The entropy of an isolated system may remain constant—or may increase. Once an egg is broken (as all of Humpty-Dumpty's horses and men discovered), it cannot be put back together again. If a container divided by an airtight partition is filled on one side of the partition with a gas and is rendered a vacuum on the other side of the partition, the result is a system with some order: gas on one side, a vacuum on the other. Withdraw the partition ever so slightly, and entropy increases as the gas comes to occupy both sides of the container indifferently. Even if you put back the partition, you cannot reduce the entropy. The order that had been represented by a gas-filled region and a region of vacuum is forever lost.

CHAPTER 10

WAVES AND SOUND

ERIODIC MOTION, MOTION THAT OCCURS AND REPEATS ITSELF, IS A PHENOMENON common in nature, most significantly in waves, and is an important theme in physics. Galileo, most familiar as an astronomer, also made fundamental observations of a simple form of periodic motion called *oscillation*. Observing the motion of a pendulum, he concluded that:

◆ The *period* of the pendulum's swing (the time duration of one cycle of the pendulum's motion) is independent of the mass of the pendulum.

This result is consistent with Galileo's finding about the behavior of free-falling objects: large and small masses fall with the same acceleration, due to gravity and independent of mass.

◆ Period is also independent of *amplitude* (the angular distance of the swing), for amplitudes of a few degrees.

On what, then, is the period dependent? The length of the pendulum and the acceleration of gravity:

$$T = 2\pi \sqrt{\frac{l}{g}}$$

Why such interest in pendulum motion?

The reason is that it is a basic type of motion—called *simple harmonic motion*—found in many physical systems. If one plots on a graph the position of the pendulum bob against time, the result is a sine or cosine function of time, which are the harmonic functions of periodic waves. Thus, pendulum motion manifests some basic properties of simple waves.

What Is a Wave?

All of us, at one time or another, have tossed a pebble into still water and watched the ripples travel out from the point of impact. These are *mechanical waves*: a disturbance in an identifiable medium (in this case, water). A remarkable property of waves is that they transmit *energy* from place to place, but the medium itself does not travel between the two places. That is, the tossed pebble causes movement within the water, but it does not cause the water itself to move from one place to another. The wave travels *through* the water. It does not *transport* the water.

Anatomy of a Wave

Waves come in various shapes, but they all have a common anatomy. They have *crests* and *troughs*, which are, respectively, the high points above the level of an undisturbed state (for example, calm water) and below it. The distance from crest to crest (or trough to trough) is the *wavelength* of the wave. The height of the wave— that is, the distance from the level of the undisturbed state to the crest of the wave— is its *amplitude*. The amount of time it takes for a wave to repeat itself at any point in space is its *wave period*. The number of wave crests that pass a given point during a given unit of time is the *frequency* of the wave. If many crests pass a point in a short period of time, we have a high-frequency wave. If few pass that point in the same amount of time, we have a low-frequency wave. Wave frequency is expressed in a unit of wave cycles per second, called the hertz (after the German physicist Heinrich Hertz), abbreviated Hz.

Wavelength and frequency are inversely related; that is, if the wavelength is doubled, the frequency is halved, and if the frequency is doubled, the wavelength is halved. Multiplying wavelength by frequency gives the wave's velocity.

In addition, waves may be *longitudinal*, *transverse*, or *mixed*.

◆ **Longitudinal polarization.** The wave moves in a direction parallel to its velocity. Sound waves traveling through liquids and gases move in this manner.

◆ **Transverse polarization.** The wave's motion is perpendicular to its velocity. This is the case with electromagnetic radiation.

◆ **Mixed waves.** Water waves behave this way, combining longitudinal and transverse motions.

Wave Interaction

In the real world, many waves interact. When two waves pass through the same point, they *interfere*.

♦ Interference may be *constructive*: The amplitudes of the two waves add to yield a larger amplitude at the point of interference.

♦ Interference may be *destructive*: The amplitudes cancel, yielding a smaller amplitude at the point of interference.

Standing waves are created when a wave interferes with itself. This is observed when a wave traveling away from its source meets its reflection traveling back.

Doppler Shifts

We are all familiar with the wah-wah change in pitch of a locomotive horn as it passes while we wait by a railroad crossing. It is not that the pitch of the horn itself (the sound source) varies, but that the sound waves are made shorter by the approach of the sound source and then longer as the sound source recedes from us (the train passes). When either the wave source or wave detector moves, changes in frequency will be detected because the movement shortens or lengthens the wave period. This is called a Doppler shift, named after the German scientist who first described it.

Sound

Some of the most familiar and easily studied waves are sound waves, which are produced by a vibrating body that compresses the air directly in front of it as it moves in one direction, then lessens the pressure on the air as it moves in the opposite direction. One such cycle of *compression* and *rarefaction* of air molecules creates one longitudinal wave.

Pitch

The *pitch* of a sound is a function of the frequency of the waves produced. High notes are produced by higher frequencies (a more rapidly vibrating body), and low notes by lower frequencies (a more slowly vibrating body).

Intensity

Intensity is related to *loudness*, but loudness is a subjective judgment made by a hearer, whereas intensity is a measurement of energy. Physicists find that the relation between intensity and loudness is nearly logarithmic. Intensity level is measured in *decibels* and is given in the equation: $\beta = 10 \log I/I_0$, where β is the intensity level of the sound in decibels, I is the sound intensity, and I_0 is the intensity of the threshold of (human) hearing (which is about 10^{-12} W/m^2).

Force Vibrations

The intensity of sound may be enhanced by *forced vibration*, as when one strikes a tuning fork and holds its base to a wooden tabletop. The vibrations of the tuning fork are transmitted to the greater surface area of the table, all of which vibrates, thereby producing a more intense sound. This is the principle behind most musical instruments, which typically use various materials of various shapes to enhance the vibration of, say, a plucked or bowed string.

Resonance

Related to forced vibration is *resonance*. If two vibrating objects are identical or if one has a natural vibration rate that is a multiple of the other's, the vibration of one will create a *sympathetic vibration*, or resonance, in the other as the sound waves it produces travel through the air and encounter the other object.

Beats

Listen to two tones that are *almost* of the same pitch (that is, almost at the same frequency). What you will hear is not two separate tones, but a tone characterized by alternating loudness, known as *beats*. The number of these beats per second equals the difference between the frequencies of the two individual waves of each tone. When a fiddler tunes his/her instrument to the pitch of a pitch pipe (for example), he/she turns the tuning peg or screw just to the point that the beats cease to be heard. This signifies that the frequencies of the two tones are now identical—the troughs and crests of the sound waves coincide perfectly.

CHAPTER 11

ELECTRICITY, MAGNETISM, AND ELECTROMAGNETISM

W E MAY THINK OF ELECTRICITY AS THE THING THAT HAPPENS WHEN WE FLIP A LIGHT switch, and we may, therefore, associate electricity with relatively recent technology. Yet, electricity was known to the ancient Greeks, who observed how small pieces of amber, when rubbed, would attract pieces of straw. The very word *electric* is derived from the Greek word for amber.

Electrostatics

As universal gravitation is a fundamental force, so are the electrical forces that attract and repel one another. The study of the behavior of such forces is called *electrostatics—static* because the electrical charges under consideration are stationary; no electric *current* is involved.

Electrical Charge

Newton's universal law of gravitation holds that masses attract one another; that is, the source of gravitational force is mass. The gravitational force of mass is always positive—and always attracts. The source of electrical force is *electrical charge*, which may be positive (protons in atomic nuclei) or negative (electrons).

◆ Charges do not appear or disappear.
◆ All electrons bear the same negative charge, wheras all protons bear the same positive charge.

◆ The presence of more electrons than protons results in a negative charge, whereas the presence of more protons than electrons creates a positive charge.

◆ In a hydrogen atom, which contains but one proton and one electron, the charges are effectively canceled out by one another, so the hydrogen atom acts as if it is without charge.

◆ Objects change their net charge by gaining or losing electrons.

Electrical Forces

The charges in atoms produce the following forces:

◆ Two objects with like charges (both positive or both negative) repel each other.

Although electric charges and forces behave similarly to gravity in many ways, in this fundamental aspect of behavior they differ. Gravitational forces *always* attract.

◆ Two objects with unlike charges (one negative, the other positive) attract each other.

In this respect, electrically charged objects behave like gravity. Coulomb's law describes the forces between two point charges, giving the magnitude of the electrostatic force (F) between them:

$$F = k\frac{Q_1 Q_2 d}{d^2}$$

where Q_1 and Q_2 are the two charges separated by distance d, and k is the proportionality constant. The SI unit of electrical charge is the *coulomb* (C), named for the eighteenth-century scientist who formulated the law also named for him. If the charge is in coulombs and the distance (d) in meters, $k = 9.0 \times 10^9$ Nm^2/C^2, and will give the force (F) in newtons.

As with gravity, electrical charge is subject to the inverse square law; that is, the force between two unlike-charged objects will decrease in proportion to the square of the distance between them. The identity that exists between electricity and gravity led Albert Einstein (and others) to believe that a single theory of particle interaction could encompass electrical, gravitational, and all other fundamental forces. A single theory to express the relationships fundamental to the universe is an as-yet unattained goal of modern physics.

Transmission of Charges

Electricity is "all around" us, just as gravity is. Indeed, the electrical attractive force between a proton and an electron is 10^{40} greater than the gravitational attractive force between these two particles. We are generally unaware of electrical forces because, in most matter, there is a balance between protons and electrons, so that most matter behaves as if it were without charge. The negatives and positives cancel each other out.

Yet, there are times when we become quite aware of electrical charge, as when, on a dry winter day, we walk across a rug, touch a metal doorknob, and get a jolt of static electricity.

Electrical charges do not pass through all material readily.

◆ *Conductors* are materials through which charges (and currents) readily pass. Metal and salt water are excellent examples of conductors.

A good conductor is a material that allows electrons to move freely. The charge is propagated along a copper wire (for example) by collisions between the electrons in the wire.

◆ *Insulators* do not readily conduct charges. Many nonmetallic materials are good insulators.

In an insulator, electrons and ions are tightly bound to fixed sites. They cannot move.

◆ *Semiconductors* are materials that have some of the properties of conductors and some of the properties of insulators. Silicon is a familiar semiconductor example.

The practical value of semiconductors is exploited in a vast array of electronic devices. Because semiconductors have fewer free electrons than conductors, they may be used to control small charges, which regulate the flow of larger charges in conductors. Thus, semiconductors function as electrical "valves" and are, therefore, invaluable in complex electronic equipment.

◆ *Superconductors* are special materials that conduct electrical charges with very great efficiency at very low temperatures.

Superconductors offer very little frictional resistance to the movement of electrons; therefore, much less energy is required to send a charge through a superconductor versus an ordinary conductor. Superconductors are used in certain specialized applications, such as ultrahigh-speed super computers.

Producing an Electrostatic Charge
Everyone is familiar with producing a static charge by rubbing an object such as a piece of hard rubber with animal fur. Rubbing such objects causes some electrons to move from the fur to the rubber, thereby negatively charging the rubber and positively charging the fur.

Once charged, the rubber holds the acquired electrons on its surface; that is, because rubber is an insulator, it does not allow the electrons to move about. If, however, the rubber is brought into contact with an uncharged object, some of the electrons will move to the uncharged object.

Contact between the charged and uncharged object is not always necessary. Even without direct contact, a charged object's electrical forces cause charges in other objects (brought near enough) to change their positions. This is called *induction*.

Electric Fields

If a small positive charge is brought into proximity with a large positive charge, the small charge "feels" a force directed away from the large charge. The data of direction and magnitude of an electrostatic force is an *electrostatic* or *electric field*: the force per unit charge exerted on a small positive test charge with a magnitude q:

$$E = \frac{F}{q}$$

You might represent an electric field in a region of space by drawing arrows at some points with lengths proportional to the magnitude of the field at those points and oriented in the proper directions (like charges repelling, unlike charges attracting). If you fill in the diagram with lines parallel to the arrows at every point, you will have a picture of the lines of force. Many lines close together indicate a strong field. In an introductory physics course, you may work with equations derived from Gauss's law to calculate electric fields.

Electrical Potential Energy

The magnitude of an electric field is the force per unit charge that field produces on a charge placed in it. What happens if you *move* the charge?

Moving a charge (Q) a distance (d) in a field (E) in a direction opposite of the field requires work: EQd. This is the amount by which the electrical potential energy of the charge increases as a result of changing the position of Q.

Electrical Potential

The change in potential energy per unit charge produced by moving Q is the change in the electrical potential between the two points separated by d. However, one can define the electric potential of a system at all points in space. *Electric potential* is a field, like E, but has a scalar rather than a vector value. Electrical potential is measured in volts (V):

$$1\,V = 1\frac{J}{C}$$

Electric Current and Electrical Circuits

An electrostatic charge is a stationary charge. Electric current is a flowing electric charge. All conductors offer friction, which resists the flow of current and can be overcome only if there is an energy source driving the flow. An element of the electric circuit must increase the potential energy of an electric charge and release it.

Current Unit

Current, the flow of an electric charge, is measured in *amperes* (A), with 1 A being the equivalent of 1 C of charge moving past a point in 1 s.

Current Flow

Electric current (measured in amperes) flows *through* a conductor or some element in the circuit *across which* there is a difference in electric potential (measured in volts).

The flow of current is subject to resistance. The more current that flows, the greater the resistance (as more electrons collide with and jostle against one another). Assuming the driving force from a source of potential difference has a fixed magnitude, current will increase only until the force of friction becomes equal to the driving force. Ohm's law shows the relationship among current (I), potential difference (voltage, V), and resistance (R):

$$I = \frac{V}{R}$$

Resistance (R) is measured in units of 1 V/A = 1 Ω (1 ohm).

Circuits

In most introductory physics courses, you will analyze (and perhaps construct) simple circuits. There are two basic kinds, *series* and *parallel.*

- ◆ *Series circuits* provide a single path for current flow. The total current from the source is the same as in any component of the circuit; the voltage across the source is the sum of the voltages across each component.
- ◆ In *parallel circuits,* the circuit components are connected across the voltage source in parallel, so that each component has the full source voltage across it. The total current delivered by the source is the sum of the currents drawn by each component.

Using Kirchhoff's laws and Ohm's law, it is possible to compute currents and voltages in any circuit.

◆ *Kirchhoff's first law* states that the sum of all currents flowing into a point in a circuit is zero. This is *conservation of charge.*

◆ *Kirchhoff's second law* states that the sum of all potential charges around a closed path in a circuit is zero. That is, each potential has one value at each point in space.

Magnetism

Magnetic forces are related to electric forces, as is evident from the fact that magnetic forces appear when electric charges are in motion. However, magnetic forces are also observed as properties of certain magnetic materials through which no electric current is moving.

Poles

Magnetic forces behave in some ways like electric forces in that they may attract or repel and they become weaker with distance. In magnets that are relatively long and thin, the behavior of the magnetic forces is as if the sources of the forces were concentrated near the ends of the magnet. These *poles* are opposites. Unlike poles attract, whereas like poles repel. Between poles, the magnetic force behaves like the electric force between two charges.

Fields

The *magnetic field* near the pole of a magnet resembles the electric field near a charge. Magnetic fields are produced by electrons spinning on their own axes while orbiting atomic nuclei. In some materials, the fields created by many spinning electrons within one atom simply cancel each other out—so the material is not perceived as magnetic. In some other materials, the fields may not be canceled within individual atoms, but the orientation of different atoms in the material is such that the fields produced by all the electrons in all the atoms cancel each other out. In materials perceived as magnetic, however, the field of many spinning and orbiting electrons add without canceling.

The lines of a magnetic field, called the *magnetic flux,* appear to start and end at the poles. In fact, this is only a function of a concept of magnetism based on the idea of poles. Magnetic flux lines are really closed loops linking loops of electric current, which produce the lines of magnetic flux.

Electromagnetism

Once it is realized that moving charges in atoms produce magnetic fields, the connection between magnetism and electricity becomes apparent. The obverse of this case is also true: moving charges in electrical circuits produce magnetic fields. The magnetic effects of electric currents are best observed in coiled-wire devices called electromagnets.

◆ *Air-core electromagnets* consist of a coil surrounding air. The current passing through each loop in turn creates fields that add together. This type of coil is called a *solenoid* and is used in many types of magnetic switching and regulating devices.

◆ *Iron-core electromagnets* consist of wire coiled around an iron core. In effect, this is a solenoid with an iron core. The iron core magnetizes when a current is passed through the coils, with the effect that its field is added to that of the coil, producing a very strong magnet.

Electromagnetic Induction

From consideration of electromagnetism, you may go on to explore *electromagnetic induction*, the electrical forces and currents in conductors that are produced by magnetic fields or by the motion of conductors through magnetic fields. The British scientist Michael Faraday observed that magnetism produces electricity. Moving a wire in a magnetic field generates a current *in* the wire—if the wire is part of a closed circuit—or a voltage (called an *induced electromotive force,* or *EMF) between the ends* of the wire—if it is not part of a closed circuit. From this, Faraday formulated a law stating that induced EMF in a coil is proportional to the product of the number of loops in the coil times the rate of change of the magnetic field linking one loop. Put another way, the law states that the EMF induced in a wire is proportional to the rate of change of the flux through the loop:

$$\varepsilon = -N\frac{\Delta\phi}{\Delta t}$$

where N is the number of loops, $\Delta\phi$ the change of flux in time, Δt. The minus sign in front of N indicates the polarity of the EMF.

It is easy to see many applications of electromagnetic induction. Much of our world today runs on it. Electric generators, electric motors, and transformers all operate on this principle.

CHAPTER 12

LIGHT AND THE ELECTROMAGNETIC SPECTRUM

JUST AS THE CONNECTION BETWEEN MAGNETISM AND ELECTRICITY IS NOT OBVIOUS from a superficial, everyday, common-sense point of view, the link between these phenomena and light is not immediately intuitive. However, all are phenomena involving *electromagnetic waves*. As we leave the topics of the previous chapter, the connection between electricity and light, as electromagnetic phenomena, becomes clearest by thinking about induction. Induction involves the transmission of energy—electromagnetic radiation—by electromagnetic waves.

What Is Light?

The short answer to the question, "What is light?" is that it is electromagnetic radiation. But, so are radio waves, microwaves, radiant heat, X rays, and so on. Yet, these are clearly different from the light that we see. How can all of these phenomena be electromagnetic radiation? What accounts for the differences between them?

The Electromagnetic Spectrum

The short answer to the last question is that the only difference between visible light and radio waves, radiant heat, X rays, and so on is the length of the wave—or the frequency of the wave, which is always inversely related to wavelength.

All forms of electromagnetic radiation—radio waves, infrared, visible light, ultraviolet, X rays, and gamma rays—may be represented on an *electromagnetic spectrum* as waves differentiated only by wavelength and frequency. Radio waves are at the low end of the spectrum, which means their waves are long and their frequency, therefore, low. Gamma rays are at the high end of the spectrum, with very short wavelengths and very high frequencies.

The radio waves your AM radio "interprets" for you are large, as high as a skyscraper, say, and about a hundred meters long (10^2 m), whereas the height of a gamma ray is that of an atomic nucleus, and its length commensurately infinitesimal. The length of a typical gamma ray wave is 10^{-14} m.

What Makes Color?

Physicists and astronomers often refer to *all* forms of electromagnetic radiation simply as "light," but what we call visible light occupies a narrow band of the electromagnetic spectrum, producing waves from 400 to 700 nanometers (nm) in length—wavelengths about the dimension of an average-sized bacterium.

Just as wavelength (or frequency) determines whether electromagnetic radiation is visible light or X rays or something else, so it determines what color we see. Our eyes respond differently to electromagnetic waves of different wavelengths. Red light, at the low end of the visible spectrum, has a wavelength of about 7.0×10^{-7} m (and a frequency of 4.3×10^{14} Hz), whereas violet light, at the high end of the visible spectrum, has a wavelength of 4.0×10^{-7} m (and a frequency of 7.5×10^{14} Hz). All of the other colors fall between these extremes, in the familiar order of the rainbow: orange just above red, yellow above orange, then green, blue, and violet.

So-called white light is a combination of all the colors of the visible spectrum. When sunlight passes through moisture droplets in the air or a glass prism, it is bent (*refracted*), splits into its component colors, and we see a rainbow, or spectrum.

Speed of Light

Light—again, this term may be taken to include electromagnetic radiation at all wavelengths—travels at very high speed: 3×10^8 m/s. (Historical attempts to measure the speed of light make for interesting reading in a physics textbook.)

Orientation of Light Waves

Light waves are transverse (see Chapter 10)—their oscillatory motion is perpendicular to the direction of the wave—and so can be *polarized*. You will doubtless see a demonstration of polarized light in lecture or work with it yourself in lab. Polarized light has vibrations confined to a single plane perpendicular to the direction of motion.

Geometrical Optics

The behavior of light was studied long before light was identified as a form of electromagnetic radiation. *Geometrical optics* is the study of how light behaves when it is *reflected* or *refracted* (bent through a prism or lens).

Reflection

Most of what we see is seen by reflected light, so the geometry of reflection is of particular importance. The original ray is the *incident ray*, and the reflection is the *reflected ray*. Perpendicular to the reflecting surface is a line called the *normal*. The angle between the incident ray and the normal is the *incident angle* (θ_i), and that between the reflected ray and the normal is the *reflected angle* (θ_r). The *law of reflection* states that $\theta_i = \theta_r$.

Although the law of reflection applies to all reflecting surfaces, not all surfaces are perfectly smooth, thereby creating *regular reflection*. Irregular surfaces create *diffuse reflection*. Diffuse reflection scatters the reflected light rays, which are nevertheless subject to the law of reflection on a local level.

You will doubtless observe and calculate reflection phenomena in *plane mirrors* and *curved mirrors*, either *concave* or *convex*.

Refraction

Refraction is the bending of light when it passes from one transparent medium into another. When an incident ray traveling through one transparent medium strikes another transparent medium, some of the light is reflected (behaving as the law of reflection predicts) and some passes through the second medium, but is refracted. The physicist Snell found a constant ratio of the sines of the angles measured from the normal to the light ray in the second transparent medium:

$$n = \frac{\sin \theta_{air}}{\sin \theta_{material}}$$

The constant n is the *index of refraction*, which depends on the optical properties of the second transparent medium. It gives a measure of the amount of bending that occurs when light travels from air into a material. Discounting the index of refraction (which varies from material to material), Snell's law is simply expressed: $n_1 \sin \theta_1 = n_2 \sin \theta_2$.

Also note that the speed of light through a vacuum is different from that through a medium (for example, it travels slower in water and even more slowly through glass). The index of refraction is also the ratio of the speed of light in a vacuum (c) and the speed of light in the medium (v):

$$n = \frac{c}{v}$$

When light travels through a medium with a higher index of refraction (n) to one with a lower n, at certain angles *all* the light is reflected. This is called *total internal reflection*. If the angle of incidence is greater than the *critical angle*, all light will reflect. If it is less, light will refract. The critical angle (θ_c) may be determined by the equation:

$$\sin \theta_c = \frac{n_2}{n_1}$$

The ability to calculate the behavior of reflected and refracted light is, of course, extremely important in designing and working with lenses and lens systems (as well as mirrors that function as lenses, as in reflecting telescopes).

Wave Optics

Newton, who did much of the fundamental work in geometric optics, thought that light consisted of streaming particles. The Dutch physicist Christian Huygens (1629–1695), however, considered light to be a wave, and he used this wave concept to explain interference and diffraction, phenomena for which geometric optics has no explanation. *Huygens's principle* states that light waves move out from a source like expanding spheres analogous to crests of water waves expanding from the impact of a dropped stone.

This concept, *wave optics*, not only accounts for reflection and refraction, but also for phenomena that cannot be explained by geometric optics.

◆ *Interference* occurs when waves from different sources interfere, either adding (constructive interference) or canceling one another out (destructive interference)

◆ *Diffraction* occurs when a wave spreads as it passes through an opening or goes around an object.

Both interference and diffraction can be demonstrated experimentally and can be calculated.

CHAPTER 13

QUANTUM PHYSICS

A S IT TURNS OUT, NEITHER CONCEPT OF LIGHT—AS A STREAM OF PARTICLES OR AS A wave—is wholly adequate to explain the behavior of electromagnetic radiation. At the beginning of the twentieth century, classical physics underwent a revolution to account for the behavior of matter on the atomic level and the behavior of matter near the speed of light.

Quantum Mechanics

Although classical physics is adequate to explain motion on the macroscopic level, *quantum mechanics* is required to analyze the behavior of molecules, atoms, and nuclei.

Black-Body Radiation

The more radiation a body absorbs, the more it must emit, provided that it is in thermal equilibrium with its surroundings. To work with equations to explain this absorption and emission, nineteenth-century statistical mechanics developed the *black body*, an idealized object that absorbs all radiation falling on it and emits all radiation that it absorbs. Using statistical mechanics, the curve of black-body radiation that results is smooth, rising higher as frequency increases, with a total emitted power that is infinite (at any temperature above absolute zero). Observation of the behavior of real-world objects that approach the properties of the idealized black body do not support these results, thereby demonstrating the inadequacy of classical physics to explain black-body radiation.

In 1900, the German physicist Max Planck offered an alternative explanation that accounted for the *observed* black-body spectrum. He concluded that matter and electromagnetic radiation at frequency f cannot arbitrarily exchange small amounts of energy, but exchange energy in fixed units (called *quanta*, hence the name *quantum mechanics*). No energy unit is smaller than the quantum. Larger amounts of energy are exchanged by adding additional quantum units. The energy in a quantum is dependent on the frequency: $E = hf$. The h is called *Planck's constant* and equals 6.626×10^{-34} Js (joule-second).

Molecules emit these quanta of energy by leaping from one energy state to another. The equation $E = hf$ is the energy of the light emitted by the jump between energy states.

In contemporary physics, quanta are called *photons*, and physicists speak of light radiation in terms of them. What Planck demonstrated, then, is that light is not a *continuous* flow of particles, nor, however, a *continuously traveling wave*, but that energy is in packets—quanta, photons—which change energy states only if the amount of energy absorbed or radiated is a discrete amount of energy. Photons, which resemble particles in their quantized, or packet, nature, also behave like waves. Thus, quantum physics revises the Newtonian idea of light as a stream of particles as well as Huygens's picture of light as waves, in effect subsuming and synthesizing the two.

Photoelectric Effect

Albert Einstein used Planck's quantum hypothesis to explain the *photoelectric effect,* the emission of electrons from some metals when light shines on the surface of the metal. Classical physics could not explain certain aspects of the photoelectric effect:

◆ The light has to be of at least a certain minimum frequency (*threshold frequency*), depending on the type of metal, to produce the photoelectric effect. Below this threshold, no photoelectric effect is produced.

◆ The kinetic energy of the photoelectrons emitted is, surprisingly, not dependent on the intensity of the light shining on the metal, an apparent violation of the law of conservation of energy.

◆ The maximum kinetic energy of the photoelectrons, however, does increase with increasing light frequency. Why should energy be dependent on frequency?

◆ The electrons are emitted almost immediately. No buildup of energy flowing into the molecules is required.

Einstein created an equation to derive the *work function* (Q), the energy required to release the electron from a particular metal: $KE_{max} = hf - Q$. He explained, point by point, the observed phenomena classical physics failed to explain:

◆ It requires a certain threshold frequency of light to provide a quanta of energy sufficient to eject an electron.

◆ The number of emitted electrons does, in fact, increase with the intensity of incident light, but their kinetic energy does not increase.

◆ An increase in frequency raises the energy level of the photons, so that the electrons ejected will have greater kinetic energy.

◆ Even at a low-light intensity, quanta may possess sufficient energy (depending on frequency) to cause the immediate emission of electrons.

Particle-Wave Duality and Its Implications

The work of Planck and Einstein showed that electrons sometimes behave like waves, and quantum objects behave like waves and particles. Because of this, physicists cannot consider *only* particle or *only* wave behavior. To explain electromagnetic radiation phenomena correctly requires *complementarity*, the simultaneous treatment of light as wave and particle.

The Bohr Atom

The British physicist Ernest Rutherford performed an experiment in 1909 demonstrating that the mass of an atom is concentrated, with a positive charge, into a small central nucleus, some 10^{-5} times smaller than the radius of the atom. The Danish physicist Niels Bohr took Rutherford's result (and other data) and combined it with the ideas of quantum mechanics to create a revolutionary new picture of the structure of the atom:

◆ Negative electrons move only in *quantized states*, certain permitted circular (Kepler) orbits around the positive nucleus under the influence of the Coulomb force.

◆ Electrons do not emit energy as long as they are in one of the allowed orbits. This is called a *stationary state*.

◆ Energy is emitted as a photon when an electron jumps from one allowed state to another. The energy of the photon equals the difference in the energies of the initial and final states, which may be expressed mathematically as $hf = E_i - E_f$.

De Broglie Waves

In 1924, the young French physicist Louis de Broglie investigated just why only certain discrete energy levels—certain permitted electron orbits—exist. He proposed that the states of the hydrogen electron are stationary waves, analogous to standing waves on a stringed musical instrument. Therefore, the circumference of an electron orbit must be an integer number of wavelengths. Thus electrons—particles— behave like waves. By showing that particles can be waves, de Broglie completed the wave-particle duality that Planck's quantum hypothesis had initiated in demonstrating that waves can be particles.

Heisenberg's Uncertainty Principle

The wave-particle duality creates a most vexing dilemma. If, at the atomic level, particles behave like waves, how is it possible to say where a particle is? In the macroscopic world, it is always possible to determine the location of an object in time and space. But, if a quantum object behaves like a wave, you cannot specify both position and momentum simultaneously or energy and time simultaneously. Thus, complete knowledge of a particle's present state and future motion is impossible for quantum objects. The more completely you determine a particle's position, the less certain you could be about its momentum, and vice versa. The uncertainty in the knowledge of the momentum multiplied by the uncertainty in the knowledge of position equals Planck's constant. In effect, Heisenberg concluded, Planck's constant represented the graininess of the universe. That is, one could not inspect the universe any more closly than the scale of Planck's constant—no more than one can enlarge a detail on a photograph that is smaller than the individual silver grains that make up the emulsion of the photographic film. The image blurs into non-information.

CHAPTER 14

NUCLEAR PHYSICS

N UCLEAR PHYSICS TAKES PHYSICS INTO THE ATOM TO MODEL THE MECHANICS OF the atomic nucleus. It was the phenomena of nuclear radioactivity, first discovered in the 1890s, that provided the keys that unlocked the nucleus.

Radioactivity

The German scientist Wilhelm Roentgen found that a beam of electrons in a vacuum, if stopped by a glass or metal wall, emitted rays that penetrated solid matter, exposed photographic film, and caused certain materials to glow. Interestingly, practical application of this discovery was almost immediately made in the realm of medicine, long before the nature of these rays—so mysterious that Roentgen dubbed them X rays—was understood.

Shortly after Roentgen's discovery, the French physicist Antoine-Henri Becquerel set about looking for natural materials that might emit X rays spontaneously. He found that uranium, thorium, and actinium emitted rays, but these were different from Roentgen's X rays.

◆ **Alpha rays.** Emitted by the heavy elements Becquerel studied, are positively charged and change direction slightly when passed through a magnetic field. Alphas are helium nuclei, two protons and two neutrons bound together. Massive, they carry sufficient energy to pass through relatively thin material barriers.

◆ **Beta rays.** Also emitted by heavy elements, are negatively charged. Passed through a magnetic field, they bend farther and in the opposite direction than the positively charged alphas. Beta rays are high-energy electrons and can penetrate more material than alphas, but would be stopped by a sheet of metal.

◆ **Gamma rays.** Photons with energy levels not only higher than alphas or betas, but higher even than X rays. Gamma rays are not deflected by passage through a magnetic field and are sufficiently energetic to penetrate many layers of most materials. A heavy element such as lead, of sufficient thickness, is required to stop gamma radiation.

Elements such as those Becquerel found to emit these rays, elements like uranium, are described as *radioactive*. It was realized that these rays are too energetic to be produced by atomic electrons, so they must come from the nuclei. In fact, radioactivity is a product of a nuclear reaction in which the original nucleus (the *parent nucleus*) decays into a *daughter nucleus* to create a new element in a process called *transmutation*.

Nuclear Structure

Atomic nuclei contain positively charged protons, with a charge opposite that of the orbiting electrons. James Chadwick demonstrated in 1932 the existence of another particle, bearing no charge, which he called the *neutron*. Together, protons and neutrons make up the structure of the nucleus and are collectively called *nucleons*.

In alpha emission, the nucleus breaks into two or more lighter nuclei. In beta emission, it is a neutron that changes, transforming into an electron and a proton, releasing energy in the process. Often, the daughter nucleus produced by alpha or beta decay is left in an *excited state*, resulting in the emission of high-energy gamma radiation.

Although the decay process is random over time in any particular nucleus, the decay of a large number of nuclei can be readily characterized statistically. Different elements have different, well-defined rates of decay; however, the decay of all elements with unstable nuclei is exponential. A certain fraction of the nuclei in a given element will decay in a given time period (say, 1 h), and the same fraction of the now-reduced number of nuclei will decay in the next minute, and so on.

Physicists characterize decay rates in terms of the *half-life of an isotope*, the amount of time it takes for half the original number of nuclei to decay. Half-lives vary widely from isotope to isotope, ranging from 10^{-6} s to more than 10^9 years.

Most elements do not decay. That is, the nuclei of most elements are not unstable. Yet, consider that a nucleus contains only neutral and positively charged particles, neutrons and protons, so it is by definition electrically unstable. What, then, holds the nucleus together?

◆ The binding force must be nonelectrical.

◆ The gravity is too weak to serve as the binding force.

◆ The binding force, whatever it is, has been named the *strong force*.

◆ The force that causes decay, a relatively rare occurrence, is called the *weak force*.

Fission

Nuclear fission was first observed by the German scientists Otto Hahn and Fritz Strassman in 1939. They bombarded uranium with neutrons for the purpose of synthesizing heavier nuclei. Instead, the reaction yielded barium. Another team of physicists, Lise Meitner and Otto Frisch, explained this surprising result as *fission*: a nuclear reaction in which a heavy nucleus breaks into two nuclei of intermediate mass and other particles. The particularly dramatic by-product of fission is the release of a very large amount of energy, 10^8 times more energy per atom than is released in ordinary chemical reactions.

In fact, the energy release from fission is in accord with the famous equation Einstein had produced in 1905 (Chapter 15):

$$E = mc^2$$

where c is the speed of light (its square is a very large number, indeed) and m is mass. The amount of energy (E) produced by a small loss of mass is tremendous because the *strong force* holding the nucleus together (the *nuclear binding force*) is so *very* strong.

Fusion

Fission, splitting heavy nuclei into lighter ones, releases large amounts of energy. *Fusion*, combining very light elements such as hydrogen and helium to produce heavier ones, releases something like five times more energy per nucleon than fission.

Fusion would be a wonderful source of energy—if a way of achieving fusion under controlled conditions could be achieved. The problem is that two nuclei separated by the distance of a few nuclear diameters repel each other by electrical forces (both contain a net positive charge) and attract only weakly because the strong force rapidly diminishes with distance. Overcoming the electrical repulsion—called the *coulomb barrier*—requires finding some way to force the nuclei together in a *controlled* manner.

Uncontrolled fusion has already been achieved. In a *fusion bomb*—commonly called a hydrogen bomb—the explosion of a small fission bomb is used as a trigger to create the tremendous degree of compression required to overcome the coulomb barrier and drive nuclei together. The result is a doomsday weapon many times more powerful than a fission bomb.

Chain Reaction

Within a very short time after the discovery of fission, Enrico Fermi, an Italian expatriate physicist working at the University of Chicago, used fission to produce the first controlled *chain reaction*. A single fission requires one neutron, which releases three product neutrons. Under certain conditions, these product neutrons could be made to collide with other nuclei, releasing more neutrons until all of the reactive material—a uranium isotope—had reacted. This is what Fermi and his colleagues accomplished in 1942.

A nuclear fission chain reaction produces a great deal of energy in the form of heat, which has been used in nuclear reactors to generate electricity (by boiling water to produce steam to drive generator turbines) and to propel large ships, including submarines and aircraft carriers.

In a reactor, the chain reaction is controlled, using a *moderator* (a substance that slows the product neutrons without absorbing them) and *control rods* (neutron-absorbing material that can be introduced to limit or stop the chain reaction). Without a moderator and control rods in the system, the energy of a chain reaction would be released as the terrible explosive force of a *fission bomb*, popularly called an atomic bomb.

More Particles

Nuclear physics is still very much an evolving discipline. Increasingly sophisticated experimental techniques have turned up particles in addition to the electron, proton, and neutron, which have given rise to a physics subdiscipline dubbed *particle physics.* Your introductory course may venture into an overview of some of the most recent work in this field.

CHAPTER 15

RELATIVITY THEORY

O NE OF THE CORNERSTONES OF CLASSICAL PHYSICS IS THE CONCEPT OF THE *INERTIAL frame of reference*. Newton's laws apply in any inertial frame—that is, any frame of reference with constant velocity. What is valid in one inertial frame must be valid in any inertial frame. A familiar illustration of inertial frames is to imagine a person on a moving train tossing a ball straight up in the air. The ball will rise and fall in a straight line—as far as the person tossing the ball is concerned. To a person observing this action from the train station, however, the ball will appear to have a parabolic trajectory, because the train is moving forward. Nevertheless, both the passenger and the stationary observer will agree that the ball obeyed the laws of mechanics. What they may not agree on is who is moving and who is standing. The train passenger may insist that he is moving while the station occupant is at rest, or perhaps he will assert the opposite, claiming that the train is at rest, and the station is moving past him. Because the frame of reference is nonaccelerating, it is impossible to prove that either frame of reference is *absolute*, or in some way preferred—the natural index against which other frames of reference must be measured.

Maxwell's Electromagnetic Equations

The nineteenth-century English physicist James Clerk Maxwell predicted that the speed of light through a vacuum is unique and constant (c), regardless of the frame of reference. Why should this be so? Classical mechanics was at a loss to provide a reason.

General among nineteenth-century physicists was a theory that, because (as they thought) all waves required some medium to travel through, and because light is a wave, therefore, light waves must be propagated in some universal substance. In the nineteenth century, this medium was called *luminous ether*. No one knew anything about this ether, except what it *had* to be:

◆ **Universal.** Present even in a vacuum and in such transparent substances as glass

◆ **Super thin.** Because the planets obviously moved through it without friction

◆ **Rigid.** So rigid that it vibrated at the extremely high frequencies of light waves

The speculation was that the ether—whatever it was—might be the *absolute* frame of reference in which the speed of light is constant.

Michelson-Morley Experiment

In the 1880s, two more scientists, Albert A. Michelson and Edward W. Morley, conducted a series of experiments with an *interferometer*, a very precise device that split a beam of light from a telescope into two paths. One path was parallel to the direction of the motion of the earth, whereas the other path was perpendicular to the first. If the ether exists, then the velocity of the earth through the ether should affect the propagation of light waves differently in the parallel and perpendicular directions. When the two images were recombined, an interference pattern of fringes should appear. Moreover, these fringes should shift if the apparatus were rotated 90°.

The fringe shift was not observed.

Special Theory of Relativity

Michelson and Morley had no explanation for the null result of their experiment. And no satisfactory explanation would come until early in the next century.

Albert Einstein did not set out to explain the Michelson-Morley experiment (in fact, he later disclaimed knowledge of it), but his 1905 *special theory of relativity* provided just such an explanation: In the first place, the luminous ether does not exist and, in the second place, there is no absolute or preferred frame of reference.

What Einstein proposed was that the laws of physics are, indeed, the same in all inertial frames, *and* the speed of light (through a vacuum) is constant regardless of the observer's frame of reference.

The first conclusion presents our common sense with no difficulty. It's easy, for example, to understand the difference between what the ball-tossing train passenger perceives and what the station occupant perceives.

It is, however, the second conclusion that defies our intuition.

Return to the train. If a person coming toward you on the train, which is moving at 10 m/s, tosses a ball toward you at 10 m/s, you would perceive that ball as coming toward you at 20 m/s. The two velocities just add. However, if that train were moving at 10 m/s or even at some fantastic velocity, say 90 percent of the speed of light, and the person shined a light at you, the light would travel toward you at c, no more, no less. Moreover, c from a source moving toward you at 90 percent of the speed of light is the same as c from a source that is stationary relative to you.

Simultaneity

Is light governed by some special laws that defy all others? Actually, it is the addition of relatively low velocities (that is, velocities that do not approach the speed of light) that represent a *special* case of Einstein's more *general* law.

And special relativity has even stranger implications.

Two events that occur simultaneously in one inertial frame are not simultaneous in another inertial frame that is moving with respect to the first. Here is an illustration of this used in many physics texts:

> You are in a transparent rocket ship. A light bulb is mounted halfway between the front and back walls of the rocket ship. The rocket passes a stationary observer. As it passes, the light bulb flashes. You, in the rocket ship, see light on the front and back walls simultaneously. The stationary observer, however, sees the light moving at c with respect to himself, regardless of direction. The rocket ship moves ahead between the time the light bulb flashes and the time the light reaches the wall. The stationary observer sees the light hit the back wall of the rocket ship before it hits the front wall.

Time Dilation

We can take the implications of relativity further. The measurements of time and distance, which *seem* absolute to us, are not, in fact, constant when compared from one inertial frame of reference with another so that the speed of light can remain a constant. Put another way: a moving clock runs slower than a stationary clock. And not just a clock. *Any* physical process, in motion, runs more slowly than the same process in a stationary situation. This phenomenon can be expressed mathematically:

$$\Delta t = \frac{\Delta t_0}{\sqrt{1 - \dfrac{v^2}{c^2}}}$$

Now, to the person who is traveling in a rocket ship, say, the clock seems perfectly normal. To an observer outside of the rocket, however, it is moving slowly. Everything is relative.

Time is not the only dimension that fails to be absolute. The *Lorentz contraction* is the effect that an observer moving with a yardstick, say, will experience. He/she will find his/her yardstick shorter compared with one held by a stationary observer. As long as neither observer compares his yardstick with that of the other, no change will be perceived. Relative to their frame of reference, reality is as harmonious with common sense as it has always been. For the person in motion, *all* distances are shortened in the direction of the motion; therefore, relative to the inertial frame, the yardstick is unchanged. The Lorentz contraction may be expressed mathematically:

$$l - l_0 \sqrt{1 - \frac{v^2}{c^2}}$$

Finally, consider the *twin paradox*. One twin in a rocket ship makes a round trip to and from a star at nearly the speed of light. The other twin stays home. When the rocket passenger returns after what seems to him a matter of days, he discovers that his twin has aged many years.

Here's the paradox: To the earth-based twin, clocks on the spaceship (if he could see them) seemed to run very slowly. Yet, to the rocket passenger, it was the *earthly* clocks that seemed to run slowly. Why, then, did the rocket passenger age only a few days, whereas his twin on earth aged years?

The difference is that the earth-based twin never left his inertial frame, whereas the rocket passenger twin accelerated at the start, middle (he made a round trip, so turned around), and end of his trip. His frame was not inertial.

Mass Increase

Special relativity predicts that the apparent mass of an object increases as its speed increases, according to a formula similar to the formulas for time dilation and for the Lorentz contraction:

$$m = \frac{m_0}{\sqrt{1 - v^2/c^2}}$$

Mass-Energy Equivalence

This is expressed in Einstein's most famous equation, $E = mc^2$. The increase of mass with speed is related to the increase in total energy with the addition of kinetic energy. A particle with mass m has a total energy of mc^2—mass times the speed of light squared, which is a very large number that represents an enormous amount of energy, if the mass could be converted to its energy (as it has been, partially, in nuclear reactors and in fission and fusion weapons).

General Relativity

Special relativity deals with the behavior of objects and events near the speed of light. *General relativity* is a theory of gravity, which explains certain effects that

Newton's theory cannot. It is not that general relativity supplants Newton, but that it shows Newtonian mechanics to be a special, as it were, local, case of general relativity.

Newton introduced the concept of gravitational force as a property of all matter possessing mass. With general relativity, Einstein proposed that matter does not merely attract matter, but that all matter *warps the space around it*. For Newton, the trajectory of an orbiting planet is curved because it is subject to the gravitational influence of, for example, the sun. For Einstein, the planet's trajectory is curved because *space itself has been curved* by the presence of the massive sun.

This change represents a fundamental shift in the way we think about the universe. One major difference between Newton's and Einstein's theories of gravitation is that if mass distorts space, then photons of light (in addition to matter) should feel the effect of gravity.

They do.

To an observer in an inertial frame of reference, light moves in a straight line. To an observer in an accelerating frame, it appears to move in a curved path. Why? Einstein concluded that light is deflected by a gravitational field. The effect of the gravitational field is the same as the effect of acceleration. That is, Einstein concluded that no experiment could be performed to prove whether you were in an accelerating rocket ship or in the gravitational field of a planet. The effects of the two are equivalent, and this is the *principle of equivalence*. Experiments performed in a uniformly accelerating frame or in an inertial frame with a gravitational field will yield the same results.

Space-Time

Ultimately, Einstein concluded, gravity creates *curved space-time*. The motion of an object could be described not in terms of the classical physics of forces, but in terms of the geometry of space itself. The source of this curvature is mass.

Black Holes

A *black hole* is the end result of the core collapse of a star that has a mass greater than three solar masses (is more than three times as massive as the sun). The black hole is so massive that it deflects photons, preventing their escape. Nothing can move faster than the speed of light. A black hole is so massive, that its *escape velocity*—the speed an object or photon must attain to escape the gravitational force of a mass—is greater than the speed of light; therefore, light itself cannot escape from a black hole.

In the vicinity of a black hole, space-time is warped such that it splits into two regions: inside and outside the black hole. What is the effect of this?

Suppose it were possible to send an indestructible probe to the event horizon of a black hole. Next, suppose we equip the probe with a transmitter broadcasting electromagnetic radiation of a known frequency. As the probe neared the event horizon, we would begin to detect longer and longer wavelengths. This shifting in wavelength is known as a *gravitational redshift*. It occurs because the photons emitted by our transmitter lose some energy in their escape from the strong gravitational field near the black hole. The reduced energy results in a frequency reduction (and, therefore, a wavelength increase) of the broadcast signals. It is called a *redshift* because, if we were looking at visible white light, it wood seem reddish, the wavelength having lengthened toward the red end of the visible spectrum.

The German astronomer Karl Schwarzschild (1873–1916) first calculated what is now called the *Schwarzschild radius* of a black hole. This value is the radius at which escape velocity would equal the speed of light (and, therefore, the radius within which escape is impossible) for a star of a given mass. Events that occur within this radius are hidden. For this reason, the Schwarzschild radius is also called the *event horizon.*

The closer our probe came to the event horizon, the greater the redshift. At the event horizon itself, the broadcast wavelength would lengthen to infinity, each photon having used all of its energy in a vain attempt to climb over the event horizon.

Suppose we also equipped the probe with a large digital clock that ticked away the seconds and that was somehow visible to us as remote observers. Because of time dilation, the clock would appear (from our perspective) to slow until the probe actually reached the event horizon, whereupon the clock—and time itself—would appear to slow to a crawl and stop. Eternity would seem to exist at the event horizon; that is, the process of falling into the black hole would appear to take forever. For, as the wavelength of the broadcast is stretched to infinite lengths, so the time between passing wave crests becomes infinitely long.

A person observing from inside the probe would perceive no changes in the wavelength of electromagnetic radiation or in the passage of time. Relative to him/her, nothing strange would be happening. Moreover (if the physical survival of the craft were possible), the passenger would have no trouble passing beyond the event horizon. To remote observers, however, the passenger would be stalled out at the edge of eternity.

Cosmology

The behavior of matter and photons near black holes provokes speculation about *cosmology*, the study of the origin, structure, and ultimate fate of the universe. Can it be that all matter in the universe curves space-time into a closed shape, like the interior of a black hole? Or, does the universe, in totality, have insufficient mass to cause such closure? Perhaps the case is somewhere between these two extremes.

It is likely that the universe expanded from a very small, very hot object in the so-called big bang about 15×10^9 years ago. Will it continue to expand forever? Or, will it, drawn by its own mass, reverse course and begin to collapse? Or, will it find another state between these two extremes? It is only because of the concept of general relativity that we are even able to ask such questions.

PART THREE

MIDTERMS
& FINALS

CHAPTER 16

UNIVERSITY OF CALIFORNIA AT BERKELEY

PHYSICS 8A: INTRODUCTORY PHYSICS: MECHANICS, WAVES, AND HEAT

Jason Zimba, Graduate Student Instructor

THREE EXAMS ARE GIVEN IN THIS INTRODUCTORY PHYSICS COURSE, TWO MIDTERMS and a final. Examples of one midterm and the final are given here.

Most of the students taking 8A at Berkeley are life science majors. I don't flatter myself that they will remember how to solve circular motion problems two years after they take my course. I am really just trying to teach them how to analyze a problem logically, and how to formulate and carry out a strategy for its solution. Life is full of word problems, and physics is the science of word problems par excellence. Most of my students will not be using much physics after they leave my classroom, so I don't expect them to remember much of it.

The course examinations are primarily a test of the students' problem-solving ability. They tend to reward those who have put in the most time and effort. In a perfect world, I would give take-home examinations, to eliminate the time factor. But there are very few schools where this is done.

The only way to learn physics is to do physics. The most important thing you can do to help yourself is to practice solving problems over and over. Just doing the homework isn't enough. Try to put each problem you see into a category: constant-acceleration, $F = ma$, energy conservation, momentum conservation, what have you. Then have a *standard approach* for each type of problem. The more practice problems you do, the more routine these standard strategies will seem.

Standard strategies work because introductory physics is not about smarts. It's about technique. Good technique will save you from making lots of mistakes. Draw clear force diagrams, make your sign conventions explicit, and so on. If you approach every problem differently, then you'll never see the big picture.

1. A spring with spring constant k and rest length l is compressed by an amount D to propel a block of mass M up a ramp.

The spring is released and the block slides up the ramp. (There is no friction anywhere.)

a. (5 pts) What is the initial acceleration of the block (just as the spring begins to decompress)?

Answer: $a_{\text{initial}} = kD/m$

When you saw the figure above and read the setup for this problem, you were probably expecting to use *conservation of energy* to solve it. We are going to get to that in part b. But part a here is trying to throw us a curve ball first, by asking a detailed question about the *motion* of the block. This calls for either *kinematic* reasoning or *force* reasoning, not energy reasoning.

A common mistake on part a would be to instinctively answer that the acceleration of the block is initially zero, because the block is not moving yet. But there are two ways to see why that's wrong.

The first way involves forces. Let's draw a force diagram for the block just as the spring begins to decompress. Neglecting the vertical forces (gravity and the normal force of the platform), the only force on the block is the spring force:

$$\boxed{\text{M}} \longrightarrow \ F_{\text{spring}}$$

sign convention:
$$\longrightarrow +$$

From the diagram it is easy to see that there is a *net horizontal force* on the block. So from Newton's second law, $F_{\text{net}} = ma$, the block *must* have some horizontal acceleration. This is the easiest way to see why the initial acceleration of the block can't be zero.

Nevertheless, it may seem confusing that the block could be accelerating even though it's not moving. But the point is that although the block is not moving *now*, it *will be moving* in a brief instant. This *change in the block's velocity* is what we mean by its *acceleration*. And that's the second way to see that the block's acceleration is nonzero: Its initial speed is *zero*, and a moment later its speed is *nonzero*. That's acceleration.

This kinematical explanation shows good understanding, and by itself it would be worth some partial credit for part a. But it would be difficult, if not impossible, to obtain the precise *value* of the initial acceleration using kinematics. To actually *calculate* the initial acceleration of the block, our only realistic approach is to go back to $F_{\text{net}} = ma$.

Looking back at the force diagram, notice that I indicated my sign convention for what counts as the positive direction. Using this sign convention for both the forces and the acceleration, F_{net}horizontal $= ma$horizontal becomes

$$+F_{spring} = m(+a)$$

F_{spring} gets a plus sign because it points to the right (our positive direction according to the diagram). The acceleration also gets a plus sign because the mass accelerates to the right.

Rearranging this to find a, we have

$$a = F_{spring}/m$$

But what is F_{spring}? Recall that the force exerted by a spring is of the form $F = -kx$. What does the x mean? It means *the amount by which the spring is compressed (or stretched) beyond its rest length*. We are given that the spring is initially compressed by an amount D beyond its rest length. So we will use D for x in the formula.

What about the minus sign in $F = -kx$? Do we want to write $a = -kD/m$ for the acceleration? No. According to the way we are using our diagram, the only time we put in signs by hand is when we are comparing the directions of things with our chosen positive direction. We have already done that for F_{spring}, so we don't want to "undo" this by putting in any more signs by hand. (This is not to mention that if we did make this mistake, then we would end up with a negative acceleration. According to our diagram, this would mean an acceleration to the left, which we know isn't correct.)

So to sum up, we have found: $a_{initial} = kD/m$.

It is worthwhile emphasizing that this expression only gives the acceleration of the mass at the *initial instant*. Why? Because kD only gives the magnitude of the *force* at that instant. As the spring extends back toward its rest length, the spring force gets smaller and smaller. Consequently, the acceleration of the mass gets smaller and smaller, too. Until finally, when the spring finally reaches its rest length, it is no longer exerting any force on the mass whatsoever. As a result, the mass is no longer accelerating either. It has reached maximum speed, and its speed will not change again until it begins to ascend the ramp.

b. (20 pts) How high above the ground does the block get before it comes to rest? Justify your approach.

Answer: $H = kD^2/2Mg$

The justification for the answer is found in the discussion that follows.

Now we get to the main thrust of the problem, which is to use conservation of energy to determine how high the block gets before it comes to rest on the ramp.

Notice that a force-then-kinematics approach will not do us any good here. The first reason is that the spring force is not a *constant* force (it is strong at first, then it weakens as the spring relaxes). This means that the acceleration of the block is *not constant* during the time when the block is in contact with the spring. (This point was emphasized in part a above.) So we will not be able to use our standard kinematics formulas to determine the maximum speed of the block.

A more serious worry is the curvature of the ramp. Because the ramp is curved in some unspecified way, there is really no hope of analyzing the motion of the block there using forces or kinematics.

In problems where the force aspects are complicated, we turn to conservation laws for help. Here we will use conservation of energy.

The problem asks us to justify our approach. What this is trying to say is, "If you are going to use conservation of energy, then you had better explain why you think that energy is conserved in the first place." A valid argument for this will have to include three pieces of information:

1. Spring forces are conservative forces.
2. There is no friction on the ramp.
3. Gravity is a conservative force.

A "conservative force" is basically a force that has a potential energy associated with it. Spring forces and gravitational forces both have potential energies associated with them, so we can use conservation of energy when they are around. By contrast, friction is a *nonconservative* force. There is no potential energy associated with frictional forces, and, in general, mechanical energy is *not* conserved in the presence of friction.

To be honest, we should probably also mention that we are neglecting air resistance. But air resistance is kind of like a friction force too, so maybe we've already taken care of that.

I approach all conservation of energy problems the same way:

1. Draw a picture of the "initial situation" and the "final situation."

2. Choose a *reference height* to serve as the zero of gravitational potential energy.

3. Write $E_{initial} = E_{final}$. This becomes $KE_{initial} + PE_{initial} = KE_{final} + PE_{final}$.

4. Look at the "initial" picture and total up all the energy you see, both kinetic and potential. This gives you the left-hand side of the equation.

5. Look at the "final" picture and total up all the energy you see, both kinetic and potential. This gives you the right-hand side of the equation.

Item number 2 is very important. Remember that the formula for gravitational potential energy is "*mgh*." But all the "*h*" means is "height above some reference level." *You* have to set the reference level, and this choice is entirely up to you. But beware: You have to use the *same reference level* for both the "initial" and the "final" situations. This is why I always recommend drawing the zero level right away, and sticking to it for the entire problem. Failure to use a consistent zero level is a very common source of errors in conservation of energy problems.

A related point I should make here is that sometimes I have students who approach different energy problems in different ways—perhaps for some problems they take the approach outlined above, while perhaps in other problems they say things like "Change in potential energy = Change in kinetic energy" or something like that. In my experience, mixing strategies like this leads to a lot of foolish errors.

So let's apply the strategy. Here is a picture of the "initial" and "final" situations.

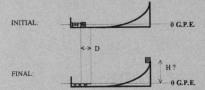

I have chosen the zero level for gravitational potential energy to be the initial height of the block. Now for the calculation:

$$E_{initial} = E_{final}$$
$$KE_{initial} + PE_{initial} = KE_{final} + PE_{final}$$

Let's look at the "initial" picture and total up all the energy we can see. Can we see any kinetic energy in the diagram? No. The block is at rest.

Can we see any potential energy in the "initial" diagram? Yes. Because the spring is compressed, there is potential energy stored in it. How much? For this we have to remember the formula for the potential energy in a spring: "$U_{spring} = (1/2)kx^2$." The "x" here is the same as the "x" in "$F = -kx$": It's the amount by which the spring is compressed beyond its equilibrium length. This is D, just as it was in part a. So the potential energy in the "initial" situation is $(1/2)kD^2$. What about gravitational potential energy in the "initial" situation? According to my diagram, the block starts off with zero gravitational potential energy. This is because in the "initial" situation, the block is located at my chosen zero level.

Now let's look at the "final" picture and total up all the energy we can see. Can we see any kinetic energy in the diagram? No. The block is again at rest.

How about potential energy? Well, in the "final" situation the spring is *relaxed*, so the potential energy that used to be there is gone. But the block is now some ways *above* my chosen zero level, which means that the block has now acquired some gravitational potential energy. The amount it has acquired is MgH, where H is the unknown height we're looking for.

So let's summarize our findings:

$$E_{initial} = E_{final}$$
$$KE_{initial} + PE_{initial} = KE_{final} + PE_{final}$$
$$0 + (1/2)kD^2 = 0 + MgH$$

Now everything is over but the algebra. We solve for H to find: $H = kD^2/2Mg$.

As an aside, what would have happened if we had chosen a *different* zero level for gravitational potential energy? Suppose, for instance, that we chose the *final* height to be the zero level, instead of the initial height. In that case, the final gravitational potential energy would have been

$$Mg(0) = 0$$

This might seem alarming, but don't forget that we also have to recalculate the *initial* potential energy. There, in addition to the $(1/2)kD^2$ worth of spring potential energy, the block also has a gravitational potential energy, equal to

$$Mg(-H) = -MgH$$

The H gets a minus sign because in the "initial" situation the block is a distance H *below* our chosen zero level. ("Up" is always positive in gravitational potential energy problems.)

So if we had set the problem up this way, we would end up with

$$KE_{initial} + PE_{initial} = KE_{final} + PE_{final}$$
$$0 + (1/2)kD^2 - MgH = 0 + 0$$

And this gives the same answer that we found above.

c. (5 pts) How much work did the spring do on the block during the launch?

Answer: $W_{spring} = (1/2)kD^2$

Many students will answer this problem as follows:

Work = force times distance

$W = (F_{spring}) \times (D)$ because the spring exerts its force over a distance D

$W = (kx) \times (D)$ because it's a spring

$W = (kD) \times (D)$ because the spring is compressed by an amount D

$$W = kD^2$$

Wrong!

Remembering that "work = force times distance" is not a bad start. But what this student forgot is that the spring force *varies* over that distance. It's only equal to kD at the *beginning* of the block's motion. Over the rest of the motion, the spring force is *smaller* than kD. So by using kD for the *whole* distance D, the student has *overestimated* the amount of work done on the block.

So are we stuck with a calculus problem here? Not if we remember the work-kinetic energy theorem.

> **Work-KE theorem.** The net work done on an object equals the change in the object's kinetic energy.

According to this theorem, if we can find the change in the block's kinetic energy, DKE, between the time when the spring starts to push and the time when the spring stops pushing, then we will also have found the net work done on the block, W_{net}. And in this problem, the only thing doing work on the block is the spring, so W_{net} is actually the quantity we're looking for.

So finding W_{spring} boils down to finding DKE for the block. We know that the *initial* kinetic energy of the block is 0, so all we have to find now is the *final* kinetic energy of the block. (Here "final" means the instant when the spring reaches its rest length.)

How can we find the final kinetic energy of the block? We might try to use kinematics to calculate the final speed v_f of the block, and then use this v_f to calculate $(1/2)Mv_f^2$. But if we try that approach, then we run into the nonconstant acceleration problem we encountered in part a.

Let's try energy conservation again. This time we'll take the "final" moment to be the moment when the spring stops pushing. The "initial" moment will be the same as in part b.

First, the diagram showing "initial" and "final":

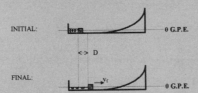

I have chosen the zero level for GPE here to be the same as in part b, but again this was arbitrary. Now let's go totaling up the energy we see:

Initial KE. Initially there is no kinetic energy, because the block is at rest.

Initial PE. Initially there is potential energy stored in the spring. Just as in part a, the spring potential energy is $(1/2)kD^2$. As for gravitational potential energy, there is none, because the block is located at the zero level.

Final KE. In the final situation, the block is moving with some (unknown) speed v_f. So the block has kinetic energy $KE_{final} = (1/2)Mv_f^2$.

Final PE. In the final situation there is no potential energy whatsoever. The spring is at its rest length, and the block is still at the zero level of gravitational potential energy.

Summing up what we've found, we have

$$KE_{initial} + PE_{initial} = KE_{final} + PE_{final}$$
$$0 + (1/2)kD^2 = (1/2)Mv_f^2 + 0$$

If necessary, we could now find the final speed of the block by solving this equation for v_f. But we may as well save the trouble, because all we are really looking for here is the final *kinetic energy* of the block. So let's back up a step or two:

$$KE_{initial} + PE_{initial} = KE_{final} + PE_{final}$$

Rearranging this, we have

$$KE_{final} - KE_{initial} = PE_{initial} - PE_{final}$$

or

$$DKE = PE_{initial} - PE_{final}$$
$$DKE = (1/2)kD^2 - 0$$

Using the work-kinetic energy theorem, $W_{net} = DKE$, we have

$$W_{net} = DKE = (1/2)kD^2$$

Because the spring is the only thing doing work on the block, the *net work* is none other than the amount of work done by the spring. Thus, $W_{spring} = (1/2)kD^2$.

This looks an awful lot like the *wrong* answer we examined above: $W = kD^2$. All that's missing is the factor of $1/2$. (Recall that the wrong answer overestimates the work done because it uses the *maximum* value for the spring force, instead of some kind of *average* value.) However, be aware that in the mind of the person grading your exam, this factor of $1/2$ is the whole ball game.

2. A spaceship of mass $5M$ is initially traveling through space in a straight line with constant speed v_0. Suddenly, there is an explosion within the ship, and the ship bursts into two pieces, one with mass $2M$ and one with mass $3M$.

   After the explosion, the $2M$ piece is traveling along the spaceship's original direction, but with speed $7v_0$.

a. (10 pts) In what direction is the $3M$ piece traveling after the explosion? Explain.
Answer: The $3M$ piece travels horizontally.

The explanation appears in the comments that follow.

An explosion is a pretty complicated thing. If we had to analyze all of the forces and accelerations that take place during an explosion, then we would be in trouble. Fortunately, there is another way out: conservation laws. When the force aspects of a problem are complicated, we turn to conservation laws for help.

But what quantity could possibly be conserved during an explosion like the one described in this problem? This is the main decision you will have to make before you can get started.

Is the *energy* of the spaceship conserved? Probably not. After all, a lot of *chemical energy* could have been released in the explosion—and if the spaceship fragments pick up any of this energy, then there will be *more* mechanical energy in the system *after* the explosion than there was *before* the explosion. So mechanical energy would certainly not be conserved in this case. (We'll check back on this in part c below.)

Is the *momentum* of the spaceship conserved? Well, let's review the condition for conservation of momentum:

In the absence of external forces, the momentum of a system is conserved.

So the thing that screws up momentum conservation is *external forces*. As long as all the forces are *internal* to the system, they don't affect momentum conservation. These forces could even be friction forces or other "dissipative" forces. Even though forces of this kind screw up *energy* conservation, they don't screw up *momentum* conservation as long as they are *internal* to the system.

So let's treat the spaceship as "the system." Are there any external forces on the spaceship? Well, because the spaceship is out in space, there's no gravity and no wind resistance. So in fact there *aren't* any external forces on the spaceship. Even the explosion itself takes place *within the system*, so it doesn't count as an external force.

Conclusion: Because there are no external forces on the spaceship, its momentum is conserved. So let's draw a picture of the initial and final situations:

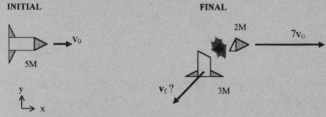

Notice that I have chosen x- and y-directions, along with definite sign conventions. (Use the *same* conventions for both the initial *and* the final.)

This figure incorporates an important part of the given information, namely that the $2M$ piece continues along the original direction with speed $7v_0$. What we are supposed to do in part a is find the direction of motion of the $3M$ piece. In the figure I just made a *guess* as to the direction of motion for this piece.

Could this guess be correct? The answer is no—but how can we see this intuitively?

Well, we have decided that the momentum of the spaceship is conserved. This means that the momentum of the system *before* the explosion is the same as the momentum of the system *after* the explosion. Keep in mind that momentum is a *vector*. It has both x- and y-components, and each component is conserved separately.

First let's look at the y-component of momentum: $p_y = mv_y$. What is the initial y-momentum of the system?

$$p_y^{\text{initial}} = mv_y^{\text{initial}}$$
$$p_y^{\text{initial}} = (5M)(0)$$
$$p_y^{\text{initial}} = 0$$

The initial y-momentum is *zero*, which is a fancy way of saying that initially, the momentum of the spaceship is purely horizontal. This should make intuitive sense based on the diagram.

Now, if the *initial* momentum is purely horizontal, then the *final* momentum must be purely horizontal too, because the initial and final momenta have to be the same.

By itself, this *doesn't* mean that both pieces of the spaceship must remain moving in the horizontal direction. After all, one piece could shoot off upwards, and the other piece could shoot off downwards, in such a way that their y-momenta *cancel out*. However, in *this problem*, we are given that the $2M$ piece moves *horizontally*. So there is no way for the y-momentum of the $2M$ piece to cancel any y-momentum in the $3M$ piece. Thus, conservation of y-momentum demands that the $3M$ piece travels horizontally.

By the way, we haven't determined whether the $3M$ piece moves backwards or forwards—only that it moves along the x-direction. We'll get down to the specifics in part b.

b. (10 pts) Find the speed of the $3M$ piece.

Answer: $v_f = -3v_0$

Well, we decided in part a that the figure we drew for the final situation couldn't really be accurate. So let's draw a better picture:

We know from our reasoning in part a that the $3M$ piece moves along the x-direction. But as of yet we don't know whether it moves backward or forward. In the figure I made a guess: It moves forward.

We can find the unknown speed v_f by requiring that the *initial* x-momentum equals the *final* x-momentum:

$$p_x^{initial} = p_x^{final}$$

The initial x-momentum is easy: it's just $(5M)(+v_0)$, with a plus sign because the velocity is in the positive x-direction, according to our diagram.

The final x-momentum has *two parts*: the x-momentum of the $2M$ piece, and the x-momentum of the $3M$ piece. The x-momentum of the $2M$ piece is $(2M)(+7v_0)$, using the given information. The plus sign is because the velocity of the $2M$ piece is in the positive x-direction, according to our diagram.

Meanwhile, the x-momentum of the $3M$ piece is $(3M)(+v_f)$, in terms of the unknown speed v_f. Again, the plus sign is because the velocity of the $3M$ piece is in the positive x-direction, according to our diagram.

So let's put these terms into the momentum conservation equation:

$$p_x^{initial} = p_x^{final}$$
$$p_x^{whole\ ship} = p_x^{2M\ piece} + p_x^{3M\ piece}$$
$$(5M)(+v_0) = (2M)(+7v_0) + (3M)(+v_f)$$

Solving for the unknown v_f gives $v_f = -3v_0$.

What does the minus sign mean? It means that the velocity of the $3M$ piece is in the *negative x-direction* according to our diagram. So it appears that our guess as to the direction of v_f was wrong. But that's okay—if you pay careful attention to the sign conventions of your diagram, then you don't have to guess right.

In retrospect, we might have made a better guess as to whether the $3M$ piece goes backwards or forward after the explosion. After all, the $2M$ piece shot ahead pretty fast. If the $3M$ piece *also* went forward, then we might have guessed that the forward momentum after the explosion would be larger than the initial forward momentum. Nevertheless, I thought it would be instructive to carry things through with the wrong guess, so you could see that it isn't crucial.

Besides, had the given information been different, things might have turned out differently. For example, if the $2M$ piece had shot ahead with a speed of only $2v_0$, then the $3M$ piece would indeed have kept moving forward. Can you calculate how fast?

c. (10 pts) How much energy was released in the explosion?

Answer: Energy released $= 60Mv_0^2$

As we discussed at the outset of part a, we don't really expect mechanical energy to be conserved during the explosion. This is because a lot of chemical energy may have been released. The interesting thing is that we can get a handle on *how much* energy was released, just by knowing how everything turned out.

Our starting point here is

$$\text{Energy released in explosion} = E_{final} - E_{initial} = (KE_{final} + PE_{final}) - (KE_{initial} + PE_{initial})$$

Notice, however, that because there are no external forces around (gravity for instance), there is no potential energy, either before or after the explosion. So we have

$$\text{Energy released} = KE_{final} - KE_{initial} = (1/2)(2M)(+7v_0)^2 + (1/2)(3M)(-3v_0)^2 + (1/2)(5M)(v_0)^2$$

or energy released = $60Mv_0^2$ in terms of given quantities.

3. A ladder of mass $3M$ leans against a frictionless wall, making an angle q with the floor. (The floor is *not* frictionless.) Suspended from the ladder, three-fourths of the way up, is a mass M.

a. (10 pts) What is the (horizontal) normal force provided by the wall?
Answer: $N_1 = (9/4)\, Mg\, \cot\theta$

This is what is known as a *statics* problem. In a statics problem, the object of interest is in a state of *equilibrium*: It has no *linear* acceleration, and it has no *rotational* acceleration. In most cases this translates into a state of *rest*, that is, a *static* state. Hence the name statics.

The approach to a statics problem is as follows:

1. Draw a force diagram for the object. *Be sure to draw the "tail" of each force vector at the place where the force actually acts.*

2. Choose a *pivot point* for calculating torques.

3. Sum up the torques acting on the object and set them equal to zero.

4. Sum up the forces acting on the object and set them equal to zero.

The results of steps 3 and 4 will yield sets of equations that you can solve for whatever quantities are unknown. In some cases, you can get the answer just by considering torques—for other problems, you will have to consider both torques *and* forces.

In this problem, the static object is the ladder. Here is a force diagram:

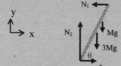

Notice that I have drawn the force vectors so that they originate at the point where they actually act. In particular, this means drawing the weight of the ladder so that it emanates from the center of the ladder. (In general, the weight acts at the "center of gravity" of the object. For a ladder with uniform mass density, the center of gravity is simply the *geometric* center.) Also notice that, as always, I have made a clear choice for the positive x- and y-directions.

The force diagram also incorporates an important piece of given information, namely that the wall is frictionless. If the wall is frictionless, then there can be no component of force parallel to the wall. The *floor*, by contrast, is *not* frictionless, so it exerts both a normal force *and* a friction force on the ladder.

How did I know which way to draw the friction force? I just used my physical intuition. It looks to me as if, were the ladder to begin slipping, it would begin to rotate *clockwise*. This means that the bottom of the ladder would slide *leftward* along the floor. The friction force opposes this incipient motion. (If you pay attention to the sign conventions in your diagram, then you shouldn't have to guess correctly here. But it's good to exercise your physical intuition anyway.)

A final point on the force diagram: In this problem, all of the force vectors are either horizontal or vertical. If, in addition, there were any *slanted* forces, then you would have to break these slanted forces into *horizontal* and *vertical* components. Having done that, it is a good idea to redraw your force diagram with the slanted forces completely replaced by their horizontal and vertical components. Things are less cluttered that way, and you are less likely to become confused.

With the force diagram in hand, we can now proceed to calculate the torques on the ladder and set the net torque equal to zero.

The first thing you have to do to calculate torques is to choose a "pivot point." The choice of pivot point is up to you, so in that sense the location is entirely arbitrary. However, as we will see, *some choices are much more convenient than others*. It takes practice to be able to choose the right pivot point in any given problem.

Once we've chosen a pivot point (we'll do this shortly), each force on the ladder contributes a torque given by

$$\tau = r_\perp F$$

Here $r_\perp$ is the perpendicular distance from *the pivot point to the line containing the force vector F.* Sometimes $r_\perp$ is called the *lever arm* or the *moment arm*.

An equivalent way of calculating the torque due to a force F is to use

$$\tau = rF_\perp$$

Here $F_\perp$ is the *component of F perpendicular to the line joining the pivot point with the point of application of the force F.*

Yet another way of calculating the torque due to a force F is to use

$$\tau = rF \sin \theta$$

Here r is the (full) distance from the pivot point to the point of application of the force, F is the (full) magnitude of the force, and q is the angle between r and F.

Of course, you don't have to memorize all three of these formulas for computing torque. Any one of them will do. They all give the same answer when applied correctly. The only thing is, sometimes one of them is easier to apply than the others, depending on the geometry of the problem at hand.

An important thing to notice is that if a force acts at the pivot point, then it contributes no torque. This is because $r = 0$ when the force acts at the pivot point. (There is no "lever arm.") You can use this fact to your advantage when choosing the pivot point, as we shall see next.

Now we're ready to start calculating torques. For reference, here's the force diagram again:

What we have to do now is choose a pivot point, calculate the torque due to each force in the diagram, and set the net torque equal to zero:

$$\tau_{net} = \tau_{N1} + \tau_{Mg} + \tau_{3Mg} + \tau_{N2} + \tau_f = 0$$

(Here each torque is labeled with the force that creates it. For example, τ_{3Mg} represents the torque due to the weight of the ladder.)

Let's choose the pivot point to be the point of contact between the ladder and the floor.

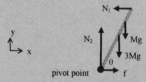

With this choice, we can now go through and calculate all the torques acting on the ladder.

First torque: τ_{N1}

To apply the formula $\tau = r_\perp F$, we have to determine the moment arm $r_\perp$ for the N_1 force. This is the perpendicular distance from the pivot point to the line containing the N_1 vector:

From the figure, $r_\perp = L\sin\theta$, where I have called the length of the ladder L. So

$$\tau_{N1} = r_\perp F = +(L\sin\theta)N_1$$

Why the plus sign? Well, we have to make a sign convention for the torques, so that some will be positive and some will be negative (otherwise, they couldn't cancel each other out). The convention I have chosen is

counterclockwise torques = positive

Perhaps you can see that the N_1 force would tend to rotate the ladder *counterclockwise* about our chosen pivot point. That's why I gave τ_{N1} a plus sign.

It's a good idea to indicate your sign convention with a symbol like this:

That way you won't forget it (and the grader will be able to understand your solution).

Second torque: τ_{Mg}

Here is the moment arm for the Mg force:

From the diagram we have

$$r_\perp = (3L/4)\cos\theta. \text{ So: } \tau_{Mg} = r_\perp F = -(3L/4)\cos\theta(Mg)$$

The minus sign is because the Mg force would tend to rotate the ladder clockwise, which counts as negative according to our convention.

Third torque: τ_{3Mg}

The trigonometry here is the same as for the Mg force; the only difference is that the point of application is $L/2$ away from the pivot point, rather than $3L/4$. So:

$$\tau_{3Mg} = r_\perp F = (L/2)\cos\theta\,(3Mg)$$

Fourth torque: τ_{N2}

Because N_2 acts at the pivot point, its moment arm is zero. So:

$$\tau_{N2} = 0$$

Fifth torque: τ_f

Likewise, because f also acts at the pivot point, its moment arm is also zero. So:

$$r_f = 0$$

Summarizing then, we have

$$\tau_{N1} + \tau_{Mg} + \tau_{3Mg} + \tau_{N2} + \tau_f = 0$$
$$= (L\sin\theta)N_1 + -(3L/4)\cos\theta\,(3Mg) + 0 + 0 = 0$$

The great thing about this is that N_1 *is the only unknown in this equation!* This happened because we chose the pivot point to be the location of two unknown forces. In general, *it's a good idea to choose the pivot point where you have unknown forces acting.* I like to think of the pivot point as "zapping" the unknown forces.

Solving the above equation for N_1, we find

$$N_1 = (9/4)\,Mg\cot\theta$$

This is the answer to part a.

b. (10 pts) What is the (horizontal) force of friction that keeps the ladder from sliding along the floor?

Answer: $f = N_1 = (9/4)\,Mg\cot\theta$

Let's look one more time at the force diagram for the ladder:

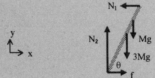

We are trying to find the friction force f, which is a *horizontal* force. So let's look at $F = ma$ in the *horizontal direction.* Remember that the ladder is in equilibrium, so the horizontal acceleration is zero. Looking at our diagram, we see that the only two horizontal forces are f and N_1. N_1 points in the negative x-direction, and f points in the positive x-direction. So, we have

$$F_{net}^{horizontal} = ma^{horizontal} = 0$$
$$-N_1 + f = 0$$

Remember that we found N_1 already in part a, by using the equation of rotational equilibrium. So the only unknown in this equation is f. Moving the N_1 to the other side, we have

$$f = N_1 = (9/4)\, Mg \cot \theta$$

(Incidentally, the fact that our answer for f is *positive* means that the friction force points to the right. Therefore, we guessed correctly as to the direction of f in the diagram.)

c. (10 pts) What is the (vertical) normal force provided by the floor?

Answer: $N_2 = 4Mg$

Let's look yet again at the force diagram for the ladder:

We are trying to find the normal force N_2, which is a *vertical* force. So, let's look at $F = ma$ in the *vertical direction*. Remember that the ladder is in equilibrium, so the vertical acceleration is zero. Looking at our diagram, we see that the vertical forces are Mg, $3Mg$, and N_2. N_2 points in the positive x-direction, and Mg and $3Mg$ point in the negative x-direction. So, we have

$$F_{net}^{vertical} = ma^{vertical} = 0$$
$$(-Mg) + (-3Mg) + (+N_2) = 0$$

This gives $N_2 = 4Mg$.

d. (10 pts) What is the smallest coefficient of friction that could make this scheme work?

Answer: $\mu_{critical} = f/N_2 = 9\cot\theta/16$

The statement of the problem emphasizes that the floor is not frictionless. And for good reason, because if *both* the wall *and* the floor were frictionless, then the ladder simply could not be in equilibrium. It would inevitably slide out of position. (This may make some intuitive sense.)

The ladder could not be in equilibrium if there were no friction between the floor and the ladder. This suggests that there is a critical amount of friction that will hold the ladder steady. Finding this critical amount is the goal of part d.

Recall that the force of static friction between two objects is related to the normal force between those two objects. Specifically,

$$f_{static\ friction} \le \mu_{static} N$$

Here μ_{static} is the "coefficient of static friction." It is determined by the kinds of *materials* that are touching one another—for example, the *wood* of the ladder and the *cement* of the floor.

Why the $\leq$ sign? The meaning of the $\leq$ sign is that *if equilibrium demands a greater friction force than* $\mu_{static} N$, *then equilibrium is simply not possible: the object will slip against the surface.*

In our problem, the static friction force is f, which we determined in part b as

$$f = (9/4)Mg\cot\theta$$

Meanwhile, in part c we determined the normal force N_2 *independently* as

$$N_2 = 4Mg$$

If the friction force f is *greater* than $\mu_{static} N_2$ where μ_{static} is the coefficient of friction between the ladder and the floor, then equilibrium is not possible. The ladder will slip. If the friction force f is *less* than $\mu_{static} N_2$, then everything is fine. Our equilibrium solution works, and the ladder stays in place.

So to find the *critical* value of the coefficient of friction, we set these two forces *equal*:

$$f = \mu_{critical} N_2$$

and solve for $\mu_{critical}$:

$$\mu_{critical} = f/N_2 = 9\cot\theta/16$$

This is the smallest value of μ_{static} consistent with equilibrium. (Note that $\mu_{critical}$ is dimensionless, as it should be.)

Notice that as the angle θ becomes *small* (corresponding to the ladder laying more and more horizontally), the required μ_{static} gets higher and higher. This is fairly intuitive: On a smooth floor, you will have to stand the ladder up more or less vertically if you want it to stay.

A crucial final note. In part b, we determined the friction force f from the requirements of equilibrium: $\tau_{net} = 0$ and $F_{net} = 0$. A *very* common mistake would be to try to find the friction force by finding N_2 first, and then setting $f = \mu_{static} N_2$. This is wrong because *we were not given that the ladder was on the verge of slipping. Only when the object is on the verge of slipping can you set* $f_{static\ friction} = \mu_{static} N$. (Remember the $\leq$ sign!) Otherwise, as in this problem, *equilibrium* determines the value of f.

<hr/>

FINAL EXAM

Practice, practice, practice. I am convinced that the more practice problems you do on your own time, the better you will do on exams.

For the exams I have written, the very best students in Berkeley's 8A course would probably answer just about everything, most of it correctly. Students scoring near the mean would answer probably three-quarters of the exam, but with substantial errors.

The mean score on 8A exams is typically between 50 and 60 percent. My exams are difficult, so I would project a mean on the low end of this range.

1. Answer each of the following questions in a sentence or two. Provide a sketch or a force diagram where appropriate.

a. (5 pts) An absent-minded physicist has left a notebook of mass M on the roof of her car. She drives in a straight line at constant speed v_0. The notebook remains in place on the roof. What is the force of static friction acting between the notebook and the car? (Neglect air resistance.)

Answer: $f_{\text{friction}} = 0$ because f_{friction} is the only horizontal force, and the notebook is not accelerating.

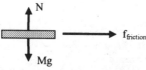

Looking at the diagram, $F = ma$ in the horizontal direction becomes $f_{\text{friction}} = Ma$. But the car is moving in a straight line at constant speed. This means that the car has no acceleration. But then the notebook has no acceleration either. So with $a = 0$ we have $f_{\text{friction}} = 0$.

You might have been tempted to write $f_{\text{friction}} = \mu N = \mu Mg$. But if this were true, then the notebook would have to be accelerating, and it's not. (Remember that $f \leq \mu N$. The only time $f = \mu N$ holds is when the object is on the verge of slipping. And why would the notebook want to slip if the car is moving in a straight line at constant speed?)

b. (5 pts) Can an object maintain a constant speed and still be accelerating? If so, then give an example. If not, then explain why not.

Answer: Yes. In uniform circular motion, the speed is constant, and the acceleration is nonzero ($a = v^2/r$).

"Constant speed" is not the same as "constant velocity." Because velocity is a *vector*, constant velocity means not only constant *speed*, but also constant *direction* (i.e., straight-line motion). If the problem had said constant *velocity*, then our answer would be that the acceleration has to be zero. (See part a for an example of this.)

c. (5 pts) Sketch the projectile's acceleration vector on the diagram below.

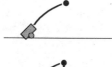

Answer:

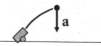

From $F_{\text{net}} = ma$, we see that the direction of the acceleration is the same as the direction of the net force. The only force is gravity, which is downward. Hence, a is downward as well.

d. (5 pts) You stand near a cliff and drop a stone over the edge. Neglecting air resistance, is the energy of the stone conserved as it falls? Why or why not? Is the momentum of the stone conserved as it falls? Why or why not?

Answer: The total energy of the stone (kinetic + potential) is conserved, because gravity is a conservative force (not a dissipative force). The momentum of the stone is *not* conserved, because an external force acts on it (namely gravity).

> Even without thinking about external forces, you can see that the momentum of the stone is not conserved as it falls, for the simple reason that the stone speeds up! (So the *y*-momentum of the stone is increasing in magnitude over time.)

2. A small block of mass M slides without friction on a curved track as shown.

The top of the track is in the shape of a circular arc with diameter D.

a. (10 pts) What *minimum* initial speed must the block have, if it is to reach the top of the hill?

Answer: $v_0 = \sqrt{2gD}$

> Because of the curvature of the track, this would be a difficult problem to solve using forces. So we will use conservation of energy. (This works because the block slides without friction.)
>
> Applying the standard conservation of energy strategy, we draw pictures of the "initial" and "final" situations.
>
>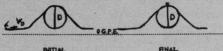
>
> Notice that, as always, I have drawn in a zero level for gravitational potential energy.
>
> In the "initial" situation, we have given the block speed v_0, which is our unknown. In the "final" situation, we have drawn the block at rest. The reason for this is that if the block has *just enough* speed to get to the top of the hill, then it will come to rest right at the top.
>
> Now that things are set up clearly, we begin totaling up the initial energy and the final energy:
>
> $$E_{\text{initial}} = E_{\text{final}}$$
> $$KE_{\text{initial}} + PE_{\text{initial}} = KE_{\text{final}} + PE_{\text{final}}$$
> $$(1/2)Mv_0^2 + 0 = 0 + MgD$$
>
> This gives
>
> $$v_0 = \sqrt{2gD}$$
>
> If you've done a lot of conservation of energy practice problems by now, then you may have jumped to this answer right away, without doing all the careful setup. However, I recommend that

you always try to use good technique on exams. You will make fewer careless mistakes that way. Keep in mind also that if you jump quickly to an answer and it's wrong, then you will get no partial credit.

b. (20 pts) What is the *maximum* initial speed the block can have, if it is not to fly off the track at the topmost point?

Answer:

$$v_{max} = \sqrt{\frac{5gD}{2}}$$

This part of the problem seems to imply that if the block is going too fast at the topmost point, then it will fly off the track. That makes sense intuitively, but it is worthwhile trying to understand it in a more careful way.

First of all, let's observe that when the block goes over the topmost point, it is moving in a curved path. But Newton's first law says that things don't like to move in curved paths. You have to *force* things to move in curved paths. (Literally!) Because of its inertia, the block would like nothing better than to fly off the track and zip out into space in a straight-line path. What could be holding the block down as it glides over the topmost point?

The answer is the block's own weight. The block's own weight is what provides the force necessary to hold it into a circular path. Try to bear this in mind as we go through the force analysis.

To get started, let's assume that when the block passes over the topmost point, it is moving at speed v_{top}. We don't know v_{top} yet, but physics is like that sometimes. To make progress, you have to put in your own stuff, and then go solve for it later if necessary.

Let's have a force diagram showing what's going on at the top of the track. In introductory physics courses, if I have one piece of advice to give you (besides "do more practice problems"), it would be this: When in doubt, draw a force diagram.

In the force diagram I have included the weight of the block as well as the normal force on the block due to the track. There are no horizontal forces whatsoever at this instant in time.

Let's sum up the vertical forces and set them equal to *ma*:

$$F_{net}^{vertical} = ma^{vertical}$$
$$N - Mg = Ma$$

Many students would go on to make a mistake at this point. Many students would set the acceleration equal to zero, because the block doesn't seem to be accelerating. (After all, its speed is constant, right?) But the block *is* accelerating. Observe that the block is moving in a *circular path*. And when an object moves in a circle at constant speed, the object is accelerating—the magnitude of the acceleration is v^2/r, and direction of the acceleration vector is toward the center of the circle.

In our problem, "toward the center of the circle" is *downward*. And downward is negative, according to the conventions in our force diagram. So our force equation becomes

$N - Mg = M(-v_{top}^2/r)$

What is r? It is the radius of the circular path. From the given information, this is $D/2$. So we have

$N - Mg = M(-v_{top}^2/(D/2)) = -2Mv_{top}^2/D$

(I set things up intentionally so that there would be a factor of 2 on the right-hand side of this equation. That way, I get to see who's really thinking and who's just plugging into standard formulas.)

The interesting thing about this equation is that *it tells me the normal force of contact, as a function of the block's speed at the topmost point*: $N = Mg - 2Mv_{top}^2/D$.

Notice what happens if the block is going too fast at the topmost point. If v_{top} is so large that $2v_{top}^2/D$ just equals Mg, then the *normal force of contact goes to zero*. This is a fancy way of saying that the *block loses contact with the track*.

So to find the critical v_{top}, we simply set the normal force equal to zero:

$$N = 0$$
$$Mg - 2v_{top}^2/D = 0$$

which gives

$$v_{top} = \sqrt{\frac{gD}{2}}$$

This is an important conclusion. But it doesn't quite answer part b. Remember, the question we have to answer is, "What is the maximum *initial* speed" so that the block doesn't fly off the track. We know how fast the block can be traveling at the *topmost* point, but we still have to relate that to the *initial* speed.

So let's go back to energy conservation. Only, this time, instead of setting things up so that the block is at rest at the topmost point, we will set things up so that the block has the speed

$$v_{top} = \sqrt{\frac{gD}{2}}$$

at the topmost point:

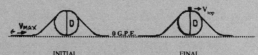

In the initial figure, v_{max} is the unknown speed we're looking for. Energy conservation gives

$$E_{initial} = E_{final}$$
$$KE_{initial} + PE_{initial} = KE_{final} + PE_{final}$$
$$(1/2)Mv_{max}^2 + 0 = (1/2)Mv_{top}^2 + MgD$$

Using our extreme value for v_{top},

$$v_{top} = \sqrt{\frac{gD}{2}}$$

this becomes

$(1/2)Mv_{max}^2 + 0 = MgD/4 + MgD$

We can solve this for v_{max}:

$$v_{max} = \sqrt{\frac{5gD}{2}}$$

This is the answer to part b.

One final note. It is fairly common to encounter a problem such as this one, where a sliding object loses contact with the surface. In all of these problems, you will need to set the normal force on the object equal to zero at some point.

3. A person is trying to tip over a heavy block of wood by pulling on it as shown.

The block is a cube of mass *M* and edge length *L*.

a. (15 pts) What is the minimum force required so that the block just begins to tip over on its corner? Answer in terms of *M*, *g*, and θ.

Answer: $T = Mg/2\cos\theta$

This is a statics problem, so the first thing to do (as with so many problems) is to draw a force diagram. Remember to draw the forces so that they originate at the point where they actually act.

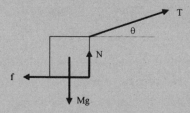

A couple of notes:

• I have drawn the normal force as acting at the *corner* of the block, because at the critical moment when the block just starts to tip, the corner is the only point of contact.

• Figuring out the direction of the friction force is a little tricky. But I can see that it must point to the left, because it has to balance the horizontal component of the tension.

It is a good idea to break the forces into perpendicular components. When this is done, we have a force diagram like so:

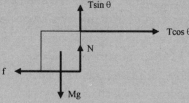

Now that we have a good force diagram, the thing to do is choose a pivot point and start calculating torques, so that we can set the net torque equal to zero. In this problem, a good choice for the pivot point is *the point about which the block actually pivots.* This is because there are two unknown forces at that point (*f* and *N*).

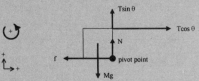

Notice that I have indicated on the diagram my sign conventions for both forces and torques. Let's set the net torque equal to zero:

$$\tau_{Mg} + \tau_{T\sin\theta} + \tau_{\cos\theta} + \tau_N + \tau_f = 0$$

The first thing we notice is that τ_N and τ_f are both zero. This is because they act at the pivot point, so the moment arm $r_\perp$ is zero.

Although it is not as obvious, the moment arm for the $\tau_{T\sin\theta}$ torque is also zero. This is because the line containing the $T\sin\theta$ vector passes through the pivot point—and this means

$$r_\perp = 0$$

This leaves τ_{Mg} and $\tau_{T\cos\theta}$ as the only nonzero torques.

From the diagram, the moment arm for τ_{Mg} is $L/2$. And the moment arm for $\tau_{T\cos\theta}$ is L. So we have (paying attention to signs)

$$(L/2)\,Mg - LT\cos\theta = 0$$

which gives

$$T = Mg/2\cos\theta$$

b. (10 pts) At what angle θ should the person hold the rope to make the block tip most easily? What is the required force in this case?

Answer: $\theta = 0$ (The person should pull horizontally.)

From our answer to part a, we can see that the required force is smallest when $\cos\theta$ is largest. And the way to make $\cos\theta$ largest is to set $\theta = 0$ (so that $\cos\theta = 1$). In other words, the person should pull *horizontally*, in which case the minimum required force is $T_{min} = Mg/2$.

c. (15 pts) The coefficient of friction between the wood block and the cement floor is $\mu_{static} = 1.5$. If the person pulls with the force calculated in part b, will the block actually tip over, or will it begin to slide instead? Explain. (If you couldn't answer part b, then use $F_{pull} = 5Mg$ as the answer to part b.)

Answer: The block will not begin to slide. The block will tip as calculated in part a. (The explanation is in the comments that follow.)

If equilibrium demands that the friction force is greater than μ_{static} times the normal force, then equilibrium is impossible: the block will slip. Conversely, if equilibrium demands a friction force that is less than or equal to μ_{static} times the normal force, then equilibrium is possible: the block will balance on its corner as calculated in part a.

So what does equilibrium demand for the friction force? Well, here's the force diagram again:

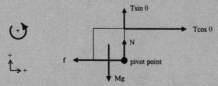

Looking at the diagram, we find from horizontal force balance that $T\cos\theta - f = 0$.

If we specialize to the case $\theta = 0$, as stated in the problem, then we have $\cos\theta = 1$ and $T = Mg/2$. So:

$$f = Mg/2$$

How does this compare with the normal force? Well, force balance in the vertical direction gives

$$T_{\sin}\theta + N - Mg = 0$$

Specializing to the case $\theta = 0$, this says

$$N = Mg$$

So we have $f = Mg/2$ and $N = Mg$. In other words, the friction force is half of the normal force. And because the friction force is less than 1.5 times the normal force, this means that the block will not begin to slide. The block will tip as calculated in part a.

4. A wood block of mass M is suspended by a cord of length L. A bullet of mass m is fired horizontally at speed v_0 and strikes the block. The block recoils from the impact, swinging out to some maximum angle θ with the bullet embedded inside.

a. (12 pts) What is the speed of the block just after the impact? Answer in terms of v_0, m, and M.

Answer: $v_{both} = mv_0/(m + M)$

Just before the collision, only the bullet is moving. Just after the collision, the bullet and block move off with some common speed v_{both}. This v_{both} is the speed we are looking for in part a.

A common mistake would be to use energy conservation here. But if you imagine what happens when a bullet slams into a wooden block, you can easily see that energy is not going to be conserved. The interior of the block splinters, the bullet deforms, and everything heats up.

That's the bad news. The good news is that all of the forces responsible for these effects are *internal* to the bullet-block system. So the *momentum* of the bullet-block system is conserved.

Let's see what momentum conservation does for us.

$$p_x^{initial} = p_x^{final}$$
$$p_x^{initial}{}_{bullet} + p_x^{initial}{}_{block} = p_x^{final}{}_{bullet} + p_x^{final}{}_{block}$$

Using the diagram, this becomes

$$m(+v_0) + 0 = m(+v_{both}) + M(+v_{both})$$

(The plus signs are because v_0 and v_{both} are to the right, which is the positive direction, according to the convention shown in our diagram.)

We solve this last equation for v_{both} as $v_{both} = mv_0/(m + M)$.

b. (18 pts) Find the maximum angle of swing θ. Answer in terms of v_0, m, M, L, and g.
Answer: $\cos \theta = 1 - m^2 v_0^2 / 2gL(m + M)^2$

After the bullet embeds itself in the block, there are no more dissipative forces, so energy is conserved. Energy conservation will give us the angle of maximum swing.

Let's have a picture of the instant just after the collision occurs, and the instant when the bullet and block reach the angle of maximum swing. To make the picture clearer, we'll change the scale of things.

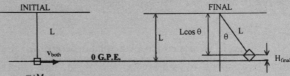

We may as well treat the bullet and block as one object of mass $(m + M)$, as shown in the diagram.

$$E_{initial} = E_{final}$$
$$KE_{initial} + PE_{initial} = KE_{final} + PE_{final}$$

Using the diagrams to total up the energy, we find

$$(1/2) (m + M) v_{both}^2 + 0 = 0 + (m + M)gH_{final}$$

Let's cancel the $(m + M)$ term from both sides:

$$(1/2) v_{both}^2 = gH_{final}$$

Before we go any further, we need to do some trigonometry to relate the final height to the angle of maximum swing. As indicated in the diagram, we have

$$H_{final} = L - L \cos \theta = L(1 - \cos \theta)$$

Together with our knowledge of v_{both} from part a, this gives

$$(1/2) m^2 v_0^2/(m + M)^2 = gL(1 - \cos \theta)$$

We can solve this for θ—or, what's simpler, $\cos\theta$. (You can always take the arccosine if you have to.)

$$\cos\theta = 1 - m^2v_0^2/2gL(m + M)^2$$

5. A block of mass $3M$ hangs from a pulley of mass M and radius R. The pulley is in the shape of a uniform disk, with moment of inertia $I = MR^2/2$.

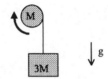

The pulley rotates without friction as the block falls under the influence of gravity.

a. (25 pts) Set up equations that determine the acceleration of the block and the tension in the rope.

Answer:

$$+3Mg - T = (3M)(+a)$$
$$T = MR\alpha/2$$
$$a = \alpha R$$

This problem combines *rotational dynamics* with *linear dynamics*. This means that in addition to using $F_{net} = ma$ for the block, we will also be using $\tau_{net} = I a$ for the pulley. Here τ_{net} is the net torque on the pulley, and a is the angular acceleration of the pulley.

The quantity I is called the *moment of inertia*, and in general it will depend not only on the pulley's mass, but also on its shape and size. In this problem the pulley is a uniform disk, which means that its moment of inertia is $MR^2/2$. Typically you would be given this formula on an exam, but check with your instructor to be sure.

As usual, the place to begin is with force diagrams. Make a separate one for each of the objects:

For the block, I have chosen the positive direction to be downward, because the block actually accelerates that way. This will help to eliminate sign errors later on.

For the pulley, I have chosen the positive sense of rotation to be clockwise. It is a very good idea to have the positive sense of rotation for the pulley to match up with the positive direction that you chose for the block. My choices match up because a positive rotation of the pulley moves the block in the positive direction.

Let's look at $F = ma$ for the block. Using the force diagram:

$$+3Mg - T = (3M)(+a)$$

Now let's look at $\tau_{net} = Ia$ for the pulley, taking torques around the center of the pulley. There are three forces acting on the pulley—its weight, the force of support from an axle, and the tension in the rope. However, *only the tension force* generates a torque around the pivot point. That's because the other two forces act right at the pivot, so their moment arms are zero.

The moment arm for the tension torque is shown on the diagram as a dotted line. This moment arm has length R. So we have

$$\tau_{net} = Ia$$
$$+(RT) = I(+\alpha)$$

The plus sign on the torque is because the tension force T tends to rotate the pulley clockwise, which is positive according to our convention. The plus sign on α is because the pulley actually accelerates in the positive sense.

If we put in the explicit moment of inertia, we get

$$RT = (MR^2/2)\alpha$$

or

$$T = MR\alpha/2$$

So far we have *two* equations, but *three* unknowns: T, a, and α. We need another equation, but where is it going to come from?

Fortunately there is a relationship between a and α, and this relationship will give us the third equation that we need. To see what the relationship is, I like to imagine that the pulley rotates by some small angle $\Delta\theta$ in a small time $\Delta\tau$. The *angular velocity* of the pulley is then

$$\omega = \Delta\theta/\Delta\tau$$

Meanwhile, by rotating by an angle $\Delta\theta$, the pulley has unwound an amount of string equal to $R\Delta\theta$. Consequently, the $3M$ mass must have dropped by this same amount. So the *linear velocity* of the $3M$ mass is

$$v = \Delta x/\Delta\tau = R\Delta\theta/\Delta\tau = R\omega$$

In other words, the *linear velocity* of the $3M$ mass is R times the *angular velocity* of the pulley. This implies that the *linear acceleration* of the $3M$ mass is R times the *angular acceleration* of the pulley. Putting this into symbols, we have

$$a = \alpha R$$

If we think of T, a, and α as our three unknowns, then this is the third and last equation that we will need to solve for them.

b. (5 pts) Solve for the acceleration and the tension.

Answer: $T = Ma/2 = 3Mg/7$

Finishing the algebra is much less important than setting up the equations, and this is reflected in the small point value for part b. You might want to skip this algebra for now and go on to another question.

Summarizing the results of part a, we found

$$3Mg - T = 3Ma$$
$$T = MR\alpha/2$$
$$a = \alpha R$$

Rearranging the third equation, we get $\alpha = a/R$. Using this in the second equation, we get $T = Ma/2$. Putting this into the first equation we get

$$3Mg - Ma/2 = 3Ma$$

which yields

$$a = 6g/7$$

and so

$$T = Ma/2 = 3Mg/7$$

6. A diver at the bottom of a lake blows an oxygen bubble with initial volume V_0.

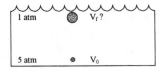

As the bubble rises to the surface, the surrounding pressure drops, causing the bubble to expand.

We will suppose that the pressure at the bottom of the lake is five atmospheres (5 atm), whereas the pressure just below the surface is 1 atm.

a. (15 pts) Assume first that the temperature of the oxygen inside the bubble remains *constant* as it rises. Under this assumption, what is the volume of the bubble just before it breaks the surface?

Answer: $V_f = 5V_0$

This is a thermodynamics problem. It is not very common to encounter thermodynamics in the first semester of an introductory physics course. So make sure that you will be responsible for this topic before you study problems like this.

If we assume that the temperature of the gas is constant during the ascent, then from the ideal gas law we have

$$p_{initial}V_{initial} = NkT_{constant}$$

and

$$p_{final}V_{final} = NkT_{constant}$$

Because $p_{initial}V_{initial}$ and $p_{final}V_{final}$ are both equal to the same thing ($NkT_{constant}$), they must be equal to each other:

$$p_{initial}V_{initial} = p_{final}V_{final}$$

This is a handy relationship for *isothermal* ideal gas processes.
Using this relationship with the given information, we have

$$(5 \text{ atm})(V_0) = (1 \text{ atm})(V_f)$$

The SI unit of pressure is the pascal, not the atmosphere. So if you want to be absolutely safe, then you should always convert atmospheres to pascals, using the conversion

$$1 \text{ atm} = 1.01 \times 10^5 \text{ Pa}$$

However, in this case, we don't really need to convert, because the units of atmospheres cancel from both sides, leaving $V_f = 5V_0$ as the final answer. The bubble has expanded fivefold in volume. (Thinking of the bubble as a sphere, notice that a fivefold increase in volume corresponds to *less* than a doubling of the diameter. Because if the diameter were to double, the volume would go up by a factor of $2^3 = 8$.)

b. (15 pts) Actually, it is more realistic to suppose that the bubble's ascent is so fast that *no heat is transferred* between the oxygen inside the bubble and the water surrounding the bubble. What is the final volume of the bubble in this case?

Answer: $V_f = 5^{3/5} V_0$

When a process allows no heat to be transferred between a gas and its surroundings, we refer to the process as *adiabatic*.
For an adiabatic ideal gas process, the initial and final pressures and volumes are related by

$$p_i V_i^\gamma = p_f V_f^\gamma$$

This looks a lot like the isothermal relationship, $p_i V_i = p_f V_f$, except for the γ exponent. This factor is called the *adiabatic constant*, and it depends on the type of particle that makes up the gas. In our case, the gas is oxygen, which is a diatomic molecule: O_2. For a diatomic molecule, the adiabatic constant is

$$\gamma = 5/3$$

Thus, we have

$$p_i V_i^{5/3} = p_f V_f^{5/3}$$

or

$$(5 \text{ atm})(V_0)^{5/3} = (1 \text{ atm})(V_f)^{5/3}$$

Again, the units of atmospheres cancel out, leaving

$$5V_0^{5/3} = V_f^{5/3}$$

To get the V_f by itself, raise each side to the 3/5 power:

$$(5V_0^{5/3})^{3/5} = (V_f^{5/3})^{3/5} = V_f$$

so that $V_f = 5^{3/5} V_0$.
In the adiabatic case, the bubble doesn't expand by as large a factor. ($5^{3/5}$ is less than 5.)

CHAPTER 17

BOWLING GREEN STATE UNIVERSITY

PHYS 201/202: COLLEGE PHYSICS

Robert I. Boughton, Professor and Chairman

USUALLY THREE MIDTERMS PLUS A FINAL ARE GIVEN IN THIS COURSE. EXAMPLES OF A midterm and the final are included here.

The primary objective of the course is to provide a survey of physics at the college algebra level. For this, students should acquire and develop skills in problem solving and simple data evaluation as well as conceptual knowledge. The examinations are problem based in such a way that all the major concepts are covered. Data handling is covered in the laboratory portion of the course. Each midterm exam is roughly one-seventh of the total grade, and the final is worth roughly two-sevenths of the total grade.

The exams included follow the general format I usually use: mostly calculation oriented. The main subject matter comes from the textbook, but I also like to include topics covered in lecture demonstrations and in the laboratory portion of the course.

SECOND MIDTERM EXAM

The best study strategy is to go over all the assigned problems and be sure you can work them (and of course understand the basic underlying concepts).

Instructions

This exam consists of five problems, each with several parts. In the calculational questions, please show your work in the space provided and write your answer in the space provided. A correct answer with no work merits *zero* points. If you run out of room, use the back side of the paper. Also note that to be complete, a physical quantity *must include* its units. A formula sheet is included at the end—tear it off and use it.

1. Concepts—*fill in the blank with light calculation*
  (12 points—equal weighting)

A 5-kg block is placed in an elevator. Consider the following three situations:

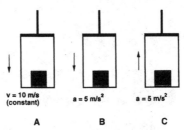

a. What is the weight of the block in each case?
  Work shown:
  $W = mg = 5$ kg $(98$ m/s$^2) = 49.0$ N
Answers:
A. 49 N
B. 49 N
C. 49 N

b. What is the normal contact force in each case?
  Work shown:

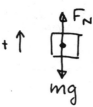

  $F_N - mg = ma$
  $F_N = m(g + a)$
  $F_N = 5$ kg$(9.8 - 5) = 5$ kg $(4.8$ m/s$^2)$
  $F_N = 5$ kg$(9.8 - 5) = 5$ kg $(14.8$ m/s$^2)$
Answers:
A. 49 N
B. 24 N
C. 74 N

The completely successful response is to provide the correct answer, including units in the blanks. Partial credit is granted for correct numerical answer but no units or incorrect units. Work that demonstrates an understanding of weight and apparent weight, force diagrams, and application of Newton's second law must be shown to gain full credit.

The major concept to understand is the distinction between real and apparent weight—that is, the modified contact force that a floor must provide if the system is accelerated, and thus what constitutes accelerated motion. The first step is to determine if the acceleration is nonzero or not. In part a, the question requires the weight itself, regardless of acceleration. In part b, the apparent weight as provided by the normal contact force is to be determined. A force diagram should be drawn and Newton's second law applied.

2. Concepts—*fill in the blank with light calculation*
(12 points—equal weighting)

A 20-newton block is acted upon by the external force indicated by the arrow in each case. The coefficient of static friction is $\mu_s = 0.9$. The coefficient of kinetic friction is $\mu_k = 0.5$.

a. What is the magnitude of the friction force in each case?
Work shown:
$f_S{}^{MAX} = \mu_s F_N$
$F_N = mg = 20 \text{ N}$
$\therefore f_S{}^{MAX} = 0.9(20 \text{ N}) = 18 \text{ N}$
Answers:
A. 5 N
B. 9 N
C. 10 N

b. What is the acceleration in each case?
Answers:
A. 0 m/s²
B. 0 m/s²
C. 0 m/s²

The completely successful response is to provide the correct answer, including units in the blanks. Partial credit is granted for a correct numerical answer but no units or incorrect units. Work that demonstrates an understanding of friction, force diagrams, and the application of Newton's second law must be shown to gain full credit.

The main concept is the relation of friction, both static and kinetic, to the normal contact force. The first step is thus to calculate the normal contact force. Then the maximum possible value of the static friction force is calculated and a determination made of whether the externally applied force can overcome it. The final conceptual understanding involves the nature of static friction—namely, that it will take on any required value to prevent motion—up to the maximum. Finally the acceleration is calculated using the second law and the net force, and is trivial in this case.

3. Calculation
(24 points)

A 1.0-kg rock on a string is observed to describe a circular path of radius equal to 0.5 m. It makes 1 revolution every 10 s.

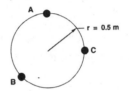

a. (5 pts) What is the speed of the rock?

Work shown:

$$v = \frac{2\pi r}{T} = \frac{2\pi (0.5 \text{ m})}{10 \text{ s}} = \frac{\pi}{10} = 0.314 \text{ m/s}$$

Answer: $v = 0.314$ m/s

b. (5 pts) What is the magnitude of the acceleration?

Work shown:

$$a_c = \frac{v^2}{r} = \frac{(0.314 \text{ m/s})^2}{0.5 \text{ m}} = 0.197 \text{ m/s}^2$$

Answer: $a = 0.197$ m/s²

c. (2 pts) What is the direction of the acceleration at point A?

Answer:

d. (2 pts) What is the direction of the acceleration at point B?

Answer:

e. (5 pts) What is the magnitude of the force acting on the rock?

Work shown:

$$F_c = ma_c = (1.0 \text{ kg})(0.197 \text{ m/s}^2) = 0.197 \text{ N}$$

Answer: $F = 0.197$ N

f. (2 pts) What is the direction of the force acting at point C?

Answer:

g. (3 pts) What agent provides this force?

Answer: Tension in string

The completely successful response is to provide the correct answer, including units in the blanks. Partial credit is granted for a correct numerical answer but no units or incorrect units. Work that demonstrates an understanding of kinematics and dynamics of uniform circular motion must be shown to gain full credit. In addition, correct conceptual understanding of the vector directions must be demonstrated.

The basic strategy is outlined in the order of the problem's parts: the speed of the object is first calculated from the circumference and the period, then the centripetal acceleration, and finally the centripetal force using the second law. The directions of these vectors are also required. A conceptual understanding that it takes some sort of agent to provide the centripetal force is also needed.

4. Calculation

(26 points)

Consider the system of two masses as shown below. The mass on the frictionless horizontal surface is $m_1 = 5.0$ kg, and the suspended mass is $m_2 = 10.0$ kg. The system starts from rest.

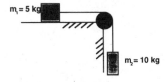

a. (6 pts) Draw the free-body diagram forces acting on each mass below. Notation: use T for tension, F_N for normal contact, and so forth.

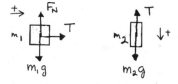

b. (6 pts) Set up the equations of motion using Newton's second law. Take the positive direction to be as indicated in the figure.

m_1:
1. $\Sigma F_x = T = m_1 a$
2. $\Sigma F_y = F_N - m_1 g = 0$

m_2: 3. $\Sigma F_x = 0$
4. $\Sigma F_y = m_2 g - T = m_2 a$

c. (6 pts) Solve the equations to find the acceleration of the system.

Work shown:

Substitute Equation 1 into Equation 4.

$$m_2 g - m_2 a = m_2 a$$
$$m_2 g = m_1 a + m_2 a = (m_1 + m_2)a$$
$$\therefore a = \frac{m_2}{(m_1 + m_2)} g = \frac{10}{15} \left(9.8 \frac{\text{m}}{\text{s}^2}\right)$$

Answer: $a = 6.5$ m/s^2

d. (6 pts) Find the tension in the cord.

Work shown:

Take the result in part c and use Equation 1:

$$T = m_1 a = (5 \text{ kg})(6.5 \text{ m/s}^2) = 32.5 \text{ N}$$

or use Equation 2:

$$T = m_2(g - a) = 10 \text{ kg}(9.8 - 6.5) = 32.5 \text{ N}$$

Answer: $T = 32.5$ N

e. (2 pts) Suppose there is kinetic friction on the horizontal surface. Indicate what will happen to the acceleration and the tension.

a: INCREASE/DECREASE

T: INCREASE/DECREASE

Work shown:

a:
$$T - f_k = m_1 a$$
$$T - \mu_k m_1 g = m_1 a$$
$$\therefore T = m_1(a + \mu_k g)$$
$$m_2 a = m_2 g - m_1 a - \mu_k m_1 g$$
$$(m_1 + m_2)a = m_2 g - \mu_k m_1 g$$
$$\therefore a = \frac{m_2}{(m_1 + m_2)} g - \mu_k \frac{m_1}{m_1 + m_2} g$$

T:

$$T = m_1 a + \mu_k m_1 g$$

$$= \frac{m_1 m_2}{(m_1 + m_2)} g - \mu_k \frac{m_1^2}{(m_1 + m_2)} g + \mu_k m_1 g$$

$$= \frac{m_1 m_2}{(m_1 + m_2)} g + \mu_k \frac{m_1^2 + m_1 m_2 - m_1^2}{(m_1 + m_2)} g$$

$$= \frac{m_1 m_2}{(m_1 + m_2)} g(1 + \mu_k)$$

Answer:

a: DECREASE

T: INCREASE

This problem includes correct drawing of force diagrams, algebraic application of Newton's law using a force diagram, and finally, calculations of numerical quantities, including units. Work demonstrating an understanding of how Newton's laws of motion are applied must be shown to gain full credit.

The strategy here is to first draw a correct force diagram, because everything else follows. In this case there are two masses to isolate and draw the forces for. Next, the algebraic application of Newton's second and third laws is used to solve for the system acceleration and cord tension from a set of simultaneous equations. Finally, a conceptual understanding of the variation of the acceleration and the cord tension if sliding friction is added is probed.

The result that the acceleration decreases can be arrived at by treating the system as a whole and determining that the net driving force is lessened.

The result that the tension increases follows from an analysis of the suspended mass, because the net force on it must be less and yet its weight has not changed.

5. Calculation

(26 points)

A 500-kg roller coaster car starts from rest at point A on the diagram. Neglect friction unless otherwise told not to.

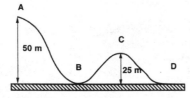

a. (6 pts) Find the total mechanical energy at point A.

Work shown:

$$E_0 = KE_0 + PE_0 = 0 + mgh_0$$
$$= (500 \text{ kg})(9.8 \text{ m/s}^2)(50 \text{ m}) = 2.45 \times 10^5 \text{ J}$$

Answer: $E_A = 2.45 \times 10^5 \text{J}$

b. (2 pts) What is the total mechanical energy at point B?

Work shown:

$$E_B = E_0 \text{ by conservation of energy}$$

Answer: $E_B = 2.45 \times 10^5$ J

c. (6 pts) Find the speed of the car at point B.

Work shown:

$$E_B = KE_B + PE_B = \frac{1}{2}mv_b^2 + 0$$

$$\therefore v_B = \sqrt{\frac{2E_B}{m}} = \sqrt{\frac{2(2.45 \times 10^5 \text{ J})}{5 \times 10^2 \text{ kg}}}$$

Answer: $v_B = 31.3$ m/s

d. (6 pts) What is the speed of the car at point C?

Work shown: by Conservation of Energy: $E_c = E_0$

$$E_c = KE_c + PE_c = \frac{1}{2}mv_c^2 + mgh_c$$

$$\therefore \frac{1}{2}mv_c^2 = E_c - mgh_c = 2.45 \times 10^5 \text{ J} - (500 \text{ kg})(9.8 \text{ m}/\text{s}^2)(25 \text{ m})$$

$$= 1.23 \times 10^5 \text{ J}$$

$$v_c = \sqrt{\frac{2E_c}{m}} = \sqrt{\frac{2(1.23 \times 10^5 \text{ J})}{5 \times 10^2 \text{ kg}}}$$

Answer: $v_C = 22.1$ m/s

e. (6 pts) Conservation of mechanical energy tells us that the car's speed is the same at point D as it is at point B. In the real world, why would you expect this *not* to be the case?

Answer: Due to friction forces doing negative work on the car, the total mechanical energy at point D will be less than at point B. Because the *PE* is zero at both points, then $KE_D < KE_B$ and, therefore, $V_D < V_B$.

This problem involves numerical application of energy concepts, as well as a short written response in the last part. The written part involves knowledge of when conservation of energy applies and what the effect of friction is on the energy balance. Work that demonstrates an understanding of kinetic and gravitational potential energy and conservation of mechanical energy must be shown to gain full credit.

A conceptual knowledge of what constitutes kinetic and potential energy, and how to calculate them at any point along the path is essential here. The application of the conservation of mechanical energy in part b and an evaluation of the total mechanical energy to determine the kinetic energy and thus the speed at that point are required for part c. A repetition of this process is called for in part d. Finally, in part e, a conceptual knowledge of when mechanical energy conservation can be applied and when it cannot be is required to correctly answer the question.

Instructions

This exam consists of six problems, each with several parts. For the calculational questions, please show your work in the space provided and write your answer in the space provided. A correct answer with no work merits *zero* points. If you run out of room, use the back side of the paper. Also note that to be complete, a physical quantity *must include* its units. A formula sheet is included at the end—tear it off and use it.

1. Concepts—*light calculation (21 points)*

A scuba diver is practicing in a large tank that is 15 m tall. His volume is 0.750 m³. Consider the three points indicated by A, B, and C, at the depths shown, in answering the following questions (2 pts each).

A 0 m

B 10 m

C 15 m

a. Where is the absolute pressure the greatest?
b. Where is the absolute pressure the smallest?
c. Where is the pressure equal to atmospheric pressure?
d. (5 pts) What is the buoyant force on the diver?

Answers:

a. C

b. A

c. A

d.
$$F_B = p_w g V = (1000 \frac{\text{kg}}{\text{m}^3})(9.8 \frac{\text{m}}{\text{s}^2})(0.75 \text{ m}^3) = 7350 \text{ N}$$

As an emergency test, holes are punched in the side of the tank at the indicated points.

e. (5 pts) What is the absolute pressure at hole B? (take $P_{\text{atm}} = 1.01 \times 10^5$ Pa)
Work shown: $P_B = P_{\text{atm}}$

Answer: 1.01×10^5 Pa

f. (5 pts) What is the speed of the effluent at each location? (Assume the speed at the top surface is approximately zero.)

Work shown:

$$\frac{1}{2}pv_B^2 + pgy_B = \frac{1}{2}pv_A^2 + pgy_A$$

$$v_A = 0$$

$$\Rightarrow v_B = \sqrt{2gd_B} = \sqrt{2(9.8\frac{m}{s^2})(10\ m)}$$

$$v_c = \sqrt{2gd_c} = \sqrt{2(9.8\frac{m}{s^2})(15\ m)}$$

Answers:

A. 0 m/s

B. 14 m/s

C. 17.1 m/s

In the first three parts, the student must give the appropriate location—no partial credit is granted. In the last three parts, the completely successful response is to provide the correct answer, including units in the blanks. Partial credit is granted for a correct numerical answer but no units or incorrect units. Work that demonstrates an understanding of the variation of pressure with depth in a fluid and Bernoulli's principle must be shown to gain full credit.

The strategy in the first three sections is to understand qualitatively the variation of fluid pressure with depth. In the second three sections, an understanding of the following fluid ideas should be tackled: how to calculate the buoyant force as equal to the weight of displaced fluid, the numerical calculation of the pressure at depth in a fluid, and the application of Bernoulli's principle to determine the velocity of effluent at different depths in the fluid calls for a conceptual understanding that the pressure is the same (atmospheric) at each opening, and that fluid flow speed increases when height above a reference decreases.

2. Calculation

(16 points)

At a temperature of 20°C, a brass doughnut [$\alpha = 19 \times 10^{-6}\ (C°)^{-1}$] has an inside hole diameter of 2.000 cm. A steel sphere [$\alpha = 12 \times 10^{-6}\ (C°)^{-1}$] has exactly the same outside diameter at this temperature, so that it does not pass through the doughnut. Answer the following questions (4 pts each).

a. For each Celsius degree of temperature increase, by how much will the doughnut hole increase in diameter?

Work shown: $d(T) = d_0 + \alpha_b d_0 \Delta T$

Change $= \Delta d \equiv d(T) - d_0 = \alpha_b d_0 \Delta T$

$= (19 \times 10^{16} C^{o-1})(2.000\ cm)(1C°) = \Delta d_{hole} = 3.8 \times 10^{-5}\ cm$

Answer: $\Delta d_{hole} = 3.8 \times 10^{-5}$ cm

b. For each Celsius degree of temperature increase, by how much will the steel sphere's diameter increase?

Work shown: $\text{Change} = \alpha_s d_0 \Delta T = (12 \times 10^{-6} \text{C}^{\circ-1})(2.000 \text{ cm})(1\text{C}^\circ)$

$$= \Delta d_{\text{sphere}} = 2.4 \times 10^{-5} \text{cm}$$

Answer: $\Delta d_{\text{sphere}} = 2.4 \times 10^{-5}$ cm

c. For each Celsius degree of increase, how much bigger is the doughnut hole diameter than the sphere diameter?

Work shown: $\Delta \equiv \Delta d_{\text{hole}} - \Delta d_{\text{sphere}} = 3.8 \times 10^{-5} \text{cm} - 2.4 \times 10^{-5} \text{cm} =$

$$\Delta d_{\text{hole}} - \Delta d_{\text{sphere}} = 1.4 \times 10^{-5} \text{cm (per C}^\circ)$$

Answer: $\Delta d_{\text{hole}} - \Delta d_{\text{sphere}} = 1.4 \times 10^{-5}$ cm (per C°)

d. How much must the temperature be raised to give a clearance fit for soldering of 0.100 mm?

Work shown: $\Delta_{\text{total}} = \Delta_{\text{per C}^\circ} (\Delta T)$

$$\Rightarrow \Delta T = \frac{\Delta_{\text{total}}}{\Delta_{\text{per C}^\circ}} = \frac{(1 \times 10^{-2} \text{cm})}{1.4 \times 10^{-5} \text{cm} / \text{C}^\circ} = \Delta T = 714 \text{C}^\circ$$

Answer: $\Delta T = 714 \text{C}^\circ$

All four sections of this problem involve calculations. The content is linear thermal expansion, and it is directly related to a demonstration performed in the lecture. For full credit, the correct answer with correct units, along with work that demonstrates understanding of thermal expansion, must be shown.

The strategy is self-explanatory; the calculations in each part lead one to the other. An understanding of linear thermal expansion and its applications to fitting of objects through holes is important here. The lecture demonstration of this very phenomenon should lead to further study.

3. Calculation *(17 points)*

A sauce pan is filled with cold water at a temperature of 10°C. The volume of water is 0.002 m³. Heat is applied at a rate of 3000 J/s (3000 W).

a. (3 pts) Determine the mass of the water.

Work shown: $m = pV = (1000 \frac{\text{kg}}{\text{m}^3})(2 \times 10^{-3} \text{m}^3) = m = 2 \text{ kg}$

Answer: $m = 2$ kg

b. (3 pts) How many moles of water are there in the pan?

Work shown:

$$n = \frac{2\text{kg}}{M.W.} = \frac{2\text{kg}}{(18 \times 10^{-3}\,\text{kg}/\text{mol})} = n = 111\text{mol}$$

Answer: $n = 111$ mol

c. (3 pts) How long does it take the water to increase in temperature by 1°C?

Work shown:

$$\Delta Q = mc\Delta T = (2\text{kg})(4186\frac{\text{J}}{\text{kg}\cdot\text{C}°})(1\text{C}°) = 8372\text{J}$$

$$\Delta t = \frac{\Delta Q}{P} = \frac{8372\text{J}}{3000\text{W}} = \Delta t = 2.8\text{s}$$

Answer: $\Delta t = 2.8$ s

d. (4 pts) How long does it take the water to rise to the boiling point?

Work shown:

$$\text{Water needs to rise by } \Delta T = 90°C$$
$$t_{\text{total}} = (90\text{C}°)\,(2.8\text{ s}) = \Delta t = 251\text{ s}$$

Answer: $t = 251$ s

e. (4 pts) After the water reaches the boiling point, how long does it take to boil the pan dry?

Work shown:

$$\Delta Q = mL_{\text{vap}}$$

$$\Delta Q = (2\text{kg})(22.6 \times 10^{5}\frac{\text{J}}{\text{kg}}) = 4.52 \times 10^{6}\,\text{J}$$

$$\Delta t = \frac{4.52 \times 10^{6}\,\text{J}}{3 \times 10^{3}\,\dfrac{\text{J}}{\text{s}}} = \Delta t = 1.51 \times 10^{3}\text{s}$$

Answer: $\Delta t = 1.51 \times 10^3$ s

All five sections of this problem involve calculations relating to fluid properties, heat, and phase transitions. To gain full credit, the correct answer with units, along with work demonstrating an understanding of density, the mole, heat capacity, and latent heat must be shown.

A clear understanding of density and the mole as a quantity of matter are basic ideas in this problem. Heat as a form of energy that produces a temperature rise in the absence of any phase change, and that causes a phase change at a fixed temperature are important concepts that should be understood here.

4. Concepts *(12 points)*

In the situations described below, indicate which of the main heat transfer processes is operative to explain the phenomenon. In some cases, there may be more than one process responsible. Use Conv, Cond, or Rad:

(2 pts each)

a. Snow melts on asphalt but collects on concrete first.
Answer: Conv

b. Thermal energy to sustain life on earth comes from the sun through a vacuum.
Answer: Rad

c. A tiled bathroom floor "feels" colder than a carpeted one, even though they're at the same temperature.
Answer: Cond

d. Having plain air between your walls is a much worse insulation design than having Styrofoam, which is mostly air.
Answer: Conv

e. A hot metal stove has a metal handle to open the oven door. The handle is made of a coiled steel wire and is not hot to the touch.
Answer: Cond

f. On a clear fall night, frost (ice crystals from water in the air) forms on grass but not on a concrete sidewalk, even though the ground is above freezing.
Answer: Rad, Cond

This problem is a conceptual test of knowledge about heat transfer processes. To get full credit, the correct answer must appear in the indicated blanks, regardless of what other choices have been made.

The student should approach questions like this one with a free-form frame of mind, so that a mental picture of the situations depicted can be formed.

5. Concepts *(10 points)*

In the following situations, a gas is observed to behave in the manner described. Determine which of the following quantities—volume, number of moles, or temperature—are being varied and in which direction: increase (inc) or decrease (dec). Use a zero to indicate no change. Assume the ideal gas law applies. (2 pts each)

a. The pressure in a tire increases as you pump it up.

Answers:

V: 0

n: inc

T: 0

b. As a helium balloon rises, the pressure decreases.

Answers:

V: inc

n: 0

T: 0

c. A rigid hot-air balloon rises and the density decreases when the burner is turned on, producing buoyancy.

Answers:

V: 0

n: dec

T: inc

d. A bubble of air rises in water and grows in size.

Answers:

V: inc

n: 0

T: 0

e. A heavyweight lies on an air mattress and it pops.

Answers:

V: dec

n: 0

T: 0

> This problem is a conceptual test of the ideal gas law. For full credit, the correct answer must be indicated. Partial credit is awarded for any one correct response in the three-blank set (V, n, T).
>
> Conceptual understanding of the relationships between the quantities in the ideal gas law is essential to answer this problem. The variations of pressure, volume, number of moles, and temperature should be well understood.

6. Calculation *(24 points)*

A thermal process involves an ideal gas working between reservoirs at $T_H = 500°C$ and $T_C = -40°C$. The process absorbs 5000 J of heat at the hot reservoir and outputs 2000 J of work. (6 pts each)

a. If this were an ideal (Carnot) heat engine, what would the efficiency be?

Work shown:

$$e_c = 1 - \frac{T_C}{T_H} = 1 = \frac{(273 + (-40))}{(273 + 500)} = 1 - \frac{233}{773} =$$

$$e_{Carnot} = 70\%$$

Answer: $e_{Carnot} = 70\%$

b. What is the difference between the heat absorbed at the hot reservoir and the heat rejected at the cold reservoir?

Work shown: $Q_H - Q_C = W = 2000$ J

Answer: $Q_H - Q_C = 2000$ J

c. What is the amount of heat rejected at the cold reservoir?

Work shown: $Q_C = Q_H - W = 5000$ J $- 2000$ J $= Q_C = 3000$ J

Answer: $Q_C = 3000$ J

d. What is the actual efficiency of the process?

Work shown:

$$e = 1 - \frac{Q_C}{Q_H} = 1 - \frac{3000J}{5000J} = e = 40\%$$

Answer: $e = 40\%$

The four parts of this problem involve calculations. The correct answer with correct units must be shown, along with work that demonstrates an understanding of heat engines and their efficiency.
   This problem involves understanding of the functioning of heat engines and the relationship between heat applied, heat rejected, and the work output. The meaning of efficiency and an understanding of the ultimate in efficiency, the Carnot engine, are required.

Useful Facts and Formulas

Conversions and Constants:

$g = 9.8$ m / s^2 = 32.2 ft / s^2
1 in = 2.54 cm   1 m = 3.281 ft   1 mi = 5280 ft   1hr = 3600 s
ρ_{water} = 1000 kg/m^3 $\rho_{mercury}$ = 13,600 kg/m^3
P_{atm} = 1.01 X 10^5 N/m^2 = 14.7 lb/in^2 = 760 mm Hg
Ideal Gas Constant: R = 8.31 J/mol•K
Avogadro's Number: N_A = 6.022 X 10^{23} particles/mol
Stefan-Boltzmann: σ = 5.67 X 10^{-8} J/s•m^2•K^4

Selected properties of water:
   Specific Heat: c_{water} = 4186 J/C°•kg
   Latent Heats: L_{fusion} = 33.5 X 10^4 J/kg $L_{vaporization}$ = 22.6 X 10^5 J/kg
   Atomic mass: 0.018 kg/mol

Formulas:

vectors:

$\vec{A} = |\vec{A}| @ \theta$ can be resolved into components by:

$$A_x = A\cos\theta \qquad A_y = A\sin\theta$$

and vice versa:

$$A = \sqrt{A_x^2 + A_y^2} \qquad \tan\theta = \frac{A_y}{A_x}$$

Fluids:

pressure: $P = \dfrac{F}{A} = -B\,\dfrac{\Delta V}{V_0}$ $\qquad\qquad$ density: $\rho = \dfrac{m}{V}$

static pressure with depth: $\Delta P = \rho\, g\, \Delta h$

Pascal's principle: change in pressure is transmitted uniformly throughout fluid

Archimedes' principle: $F_B = \rho_{fluid}\, g\, V_{object}$ (weight of fluid displaced)

Bernoulli's principle:

$$P_A + \frac{1}{2}\rho\, v_A^2 + \rho\, g\, h_A = P_B + \frac{1}{2}\rho\, v_B^2 + \rho\, g\, h_B$$

Equation of continuity:

$$\rho_1 A_1 v_1 = \rho_2 A_2 v_2 \qquad\qquad \text{volume flow rate: } Q = A\, v$$

Thermal Expansion:

linear: $L = L_o + \alpha\, L_o\, \Delta T$

areal: $A = A_o + 2\alpha\, A_o\, \Delta T$

volume or bulk: $V = V_o + 3\alpha\, V_o\, \Delta T = V_o + \beta\, V_o\, \Delta T$

Heat Capacity:

$$Q = m\, c\, \Delta T \qquad c = \text{specific heat of substance}$$

Latent Heat :

$$Q = m\, L \qquad L = \text{latent heat associated with the process: fusion, vaporization, sublimation}$$

Heat Transfer:

Thermal Conduction: $\quad Q = k\, A\, \Delta T\, t / L$
k = thermal conductivity
A = cross-sectional area
L = path distance of the heat flow

Thermal Radiation: $\quad Q_{net} = \sigma\, e\, A\, t\, (T^4 - T_o^4)$

e = emissivity
A = surface area of radiator or absorber

Ideal Gas:

Ideal gas law: $P V = n R T$

R = Ideal Gas Constant

n = number of moles = mass/atomic mass

Thermodynamics:

First Law: $\Delta U = U_f - U_i = Q - W$

Q = heat input to system

W = work output by system

U = internal energy

Isothermal gas process:

$$W = n R T ln\left(\frac{V_f}{V_i}\right)$$

Adiabatic gas process:

$$W = 3/2 \; n R (T_f - T_i)$$

Heat Engines:

$W = Q_H - Q_C$

efficiency = e = $W/Q_H = 1 - Q_C/Q_H$

Carnot engine: $\dfrac{Q_C}{Q_H} = \dfrac{T_C}{T_H}$

Carnot efficiency: $e_{Carnot} = 1 - \dfrac{T_C}{T_H}$

CHAPTER 18

PHYSICS 10B: INTRODUCTION TO THE LIFE SCIENCES II

Craig Blocker, Professor

TWO MIDTERMS AND ONE COMPREHENSIVE FINAL ARE GIVEN IN THIS COURSE. AN example of a midterm and the final are presented here.

My first objective is that the students understand the basic laws of physics that are covered. This means that they not should just be able to write down equations, but that they should understand what the equation is actually saying and how to apply that equation to a problem. I am not interested in the students memorizing equations; they can always look them up if they need them later. I am more interested in their understanding the equations. For this reason, I always give the students the relevant equations on exams, so they don't have to spend time memorizing them.

A second objective is to give the students a sense that nature is described accurately by quantitative laws.

A third objective is to increase the problem-solving skills of the students, particularly for quantitative problems in which the correct answer must be deduced by reasoning from one or more basic principles.

I want my students to gain an understanding of the basic laws of physics and be able to apply these laws to solve problems. The main obstacle to overcome here is the students' tendencies to jump to a conclusion without carefully thinking through and applying the relevant physics.

The exams that I give have problems to be solved that are similar to homework problems. These problems usually are not ones in which the information given is simply plugged into a formula. Also, I do not give exams that directly test physics concepts (such as some multiple-choice and essay questions). I have found that students often can quote the concepts fine, if they have been taught that way, but that they have difficulty applying these concepts to problems. I am interested in physics being a tool that the student can use to understand problems they encounter in the future studies and careers.

EXAM 2

Be sure to write your name on your blue book. Do all your work in your blue book. Credit will be given for correct answers *only if* the work supports it. No books, papers, notes, or calculators are allowed for this exam. Please do any simple arithmetic (such as $2 \times 3 = 6$) and leave any complex arithmetic as a calculation [such as $(8.2 \times 10^8)(4.39)$]. If anything is unclear about the exam, please ask me. There are five problems on four pages of the exam. Each problem is worth 20 points, for a total of 100 points.

1. Two positive charges of 1-μC each are placed at -1 m and +1 m on the *x* axis and are fixed. A third charge of 2-μC is placed at +2 m on the *y* axis.

   a. What is the electric potential energy of the 2-μC charge?
   b. What is the electric potential due to the 1-μC charges at the point $x = 0, y = 2$ m?
   c. The 2-μC charge has a mass of 5 kg and is released at rest at the point $x = 0$, $y = 2$ m. What is the velocity of the charge when it reaches a distance of 4 m from the origin?

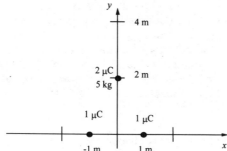

Answers:

a. The potential energy of the 2-μC charge is the sum of the potential energies due to each of the other charges.

$$U = \sum \frac{qq'}{4\pi\varepsilon_0 r} = 2\frac{9 \times 10^9 \, \text{Nm}^2 C^{-2} (1\mu C)(2\mu C)}{\sqrt{(1 \, \text{m})^2 + (2 \, \text{m})^2}}$$

b. Electric potential is electric potential energy per unit charge ($V = U/q'$), giving

$$V = \frac{U}{q'} = \sum \frac{q}{4\pi\varepsilon_0 r} = 2\frac{9 \times 10^9\,\text{Nm}^2\text{C}^{-2}(1\mu\text{C})}{\sqrt{(1\,\text{m})^2 + (2\,\text{m})^2}}$$

c. By conservation of energy, the total energy when the particle is released is equal to the total energy when it reaches $y = 4$ m. When it is released, the kinetic energy is zero (because it is released at rest). Thus, we have

$$U(y = 2\,\text{m}) = \tfrac{1}{2}mv^2 + U(y = 4\,\text{m})$$

$$\Rightarrow v = \sqrt{\frac{2\big(U(2\,\text{m}) - U(4\,\text{m})\big)}{m}}$$

$$= \sqrt{\frac{2}{5\,\text{kg}}\left[2\frac{9 \times 10^9\,Nm^2C^{-2}(1\mu C)(2\mu C)}{\sqrt{(1\,\text{m})^2 + (2\,\text{m})^2}} - 2\frac{9 \times 10^9\,Nm^2C^{-2}(1\mu C)(2\mu C)}{\sqrt{(1\,\text{m})^2 + (4\,\text{m})^2}}\right]}$$

a. For this part of the problem, it is important to first identify the formula for potential energy of a point charge. Next, it is necessary to realize that the total potential energy is the sum of that from each point charge. Finally, the student must remember that potential energy is a scalar quantity (that is, not a vector) and so the potential energies of the two 1-μC charges add algebraically.

Elements of a correct answer are (1) the correct potential energy formula and (2) the correct algebraic addition of the potential energies of the 1-μC charges, including the correct calculation of the distance.

b. This part of the question probes whether the student understands the concept of electric potential as potential energy per unit charge. Then the calculation is straightforward. It is also possible to solve the problem by using the formula for the electric potential of a point charge. Again, the correct superposition of the scalar electric potential must be understood.

c. This part of the problem requires the student to first understand that there is a force on the 2-μC charge due the electric fields of the 1-μC charges and by symmetry this force points in the y direction. It is possible to calculate this force from Coulomb's law and then get the acceleration from Newton's second law. However, this acceleration is not constant (it depends on the distance from the 1-μC charges). Because this course doesn't use calculus and because the acceleration is not constant, students cannot employ acceleration methods to solve the problem. The student must realize (hopefully guided by part a) that energy methods are necessary and understand that, because this is a problem involving electric forces (which are conservative), that energy is conserved. This leads directly to the solution above.

2. A parallel plate capacitor consists of two plates, each with area A, separated by a distance d. The capacitor is connected to a battery that charges it to a charge Q. Next, the battery is disconnected. Then the plates are moved to a distance $2d$ apart.

a. By what factor (ratio) does the capacitance change?
b. By what factor (ratio) does the charge on the plates change?
c. By what factor (ratio) does the electric potential between the plates change?
d. By what factor (ratio) does the energy stored change?

e. Where did the energy in part d come from or go to?

Answers:

a. Because the capacitance is $C = \varepsilon_0 A/d$, doubling the separation d will halve the capacitance; that is, the factor is 1/2.

b. Because the battery is disconnected, the charge has nowhere to flow, and hence doesn't change; that is, that factor is 1.

c. The potential is related to the capacitance and charge by $Q = CV$; that is, $V = Q/C$. Because the charge doesn't change and because the capacitance is halved, the potential is doubled; that is, the factor is 2.

d. The stored energy is $U = Q^2/2C$. Because the charge is unchanged and the capacitance is halved, the stored energy is doubled; that is, the factor is 2.

e. Because the plates have opposite charges, they attract each other. Thus, it is necessary to exert a force to pull them apart. The work done by this force supplies the additional stored energy in the capacitor.

This problem tests the student's understanding of simple scaling and how capacitance depends on various parameters.

a. For this part of the problem, it is first necessary to understand how capacitance depends on the separation of the plates. Because the capacitance varies as the inverse of the separation, doubling the separation halves the capacitance. Because capacitance depends only on geometric quantities (area A and separation d) and a fundamental constant (ε_0), this part of the answer does not depend on the battery or how it was manipulated.

b. Here it is necessary to understand that if the battery is disconnected, there is no path for the charge to flow through. Because the charge is conserved and has nowhere to go, the charge on the capacitor cannot change.

c. Here the student should realize that, because the battery is disconnected, the electric potential across the capacitor can change. Next, the student should remember that the potential across a capacitor, the charge on the capacitor, and the capacitance are always related by $Q = CV$. Knowing how C and Q change from parts a and b, the change in V can be determined.

d. The stored energy can be written in several ways to express the stored energy of a capacitor that depends on Q, V, and C. Any of these can be used; however, it is simpler to choose one that involves Q, because Q doesn't change. Thus, either $U = QV/2$ or $U = Q^2/(2C)$ work well and yield the same answer.

e. Answering this question relies on the student recognizing that there are no nonconservative forces, so that any change in the energy of the system must be due to work done on the system. Further, the student should realize that the plates attract each other (they have opposite charges on them); hence, a force must be exerted to pull them apart.

---

3. A 4-V battery is connected to a network of three resistors having resistances of $4\,\Omega$, $3\,\Omega$, and $1\,\Omega$ as shown below. (Note: the numbers and calculations in this problem are simple enough that you should get actual numerical answers.)

a. What is the equivalent resistance of the three resistors?

b. What is the current flowing through the battery?

c. What is the current flowing through each resistor?

d. What is the power dissipated in each resistor?

e. What is the power supplied by the battery?

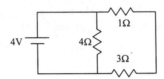

Answers:

a. The 1-Ω and 3-Ω resistors are in series, so their resistances add, giving 4 Ω. This 4 Ω is in parallel with the 4-Ω resistor, so that their resistances are combined as

$$\frac{1}{R_{eq}} = \frac{1}{R_1} + \frac{1}{R_2} = \frac{1}{4\ \Omega} + \frac{1}{4\ \Omega} = \frac{1}{2\ \Omega} \Rightarrow R_{eq} = 2\ \Omega$$

Thus, the equivalent resistance is 2 Ω.

b. The current flowing through the battery is equal to the current flowing through the equivalent resistance, that is, $I = V/R_{eq} = 4\ \text{V}/2\ \Omega = 2\ \text{A}$.

c. The 4-Ω resistor has 4 V across it, so its current is 4 V/4 Ω = 1 A. The 1-Ω and 3-Ω resistors have 4 V across their equivalent resistance of 4 Ω. Because they are in series, they have the same current and they are equal to 4 V/4 Ω = 1 A.

d. The power dissipated in a resistor is equal to I^2R. For the 4-Ω resistor, we have $P = (1\ \text{A})^2\,4\ \Omega = 4\ \Omega$. For the 1-$\Omega$ resistor, we have $P = (1\ \text{A})^2\,1\ \Omega = 1\ \Omega$. For the 1-$\Omega$ resistor, we have $P = (1\ \text{A})^2\,3\ \Omega = 3\ \Omega$.

e. The power supplied by a battery is $P = IV = 2\ \text{A}(4\ \text{V}) = 8\ \Omega$. Note that this is also the sum of the powers dissipated in the three resistors.

---

a. In this part, the first step is to recognize that the 1-Ω and 3-Ω resistors are in series. This is true in spite of the fact that they look to be in parallel because they are connected on one end, not connected on the other end, and nothing else is connected to the common end, which is the actual definition of a series connection. Thus, the 1-Ω and 3-Ω resistors have an equivalent resistance of 4 Ω. Now, this equivalent resistance is in parallel with the 4-Ω resistor (both ends of both resistors are connected together). Thus, the equivalence resistance is the parallel combination of these two 4-Ω resistors.

b. The student must realize that the definition of equivalent resistance means that the same current flows in the equivalent resistor as it does in the resistor network for the same voltage applied.

Because the equivalent resistance was calculated in part a, it is straightforward to calculate the current through the equivalent resistor (and hence through the battery) by Ohm's law.

c. For the 4-Ω resistor, the student should recognize the voltage across the resistor is equal to the voltage across the battery because potential difference is path-independent. Thus by Ohm's law (which always relates the current, voltage, and resistance of a resistor) the current can be determined. This same reasoning applies to the 4-Ω equivalent resistance of the 1-Ω and 3-Ω resistors. Because charge is conserved and doesn't accumulate at points in a circuit, the current that flows in each resistor of a series combination is the same and is equal to the current in the equivalent resistor.

d. The power dissipated in a resistor is given by several formulas involving various combinations of I, V, and R. Because we know R and I for each resistor, the formula $P = I^2R$ is the easiest formula to apply.

e. This problem can be solved in either of two ways. First, the energy supplied by a battery is IV, which can be calculated from the voltage of the battery and the current through it from part b. Alternately, because the only place for power supplied by the battery to go is dissipation by the resistors, the power supplied by the battery is equal to the sum of the powers dissipated by the resistors as determined in part d.

4. A 5-V battery with internal resistance 1 Ω is connected in parallel with a 10-V battery with internal resistance 2 Ω, and then the two batteries are connected to a 10-Ω resistor (see the following diagram). Let I_1 be the current in the 1-Ω resistor, I_2 be the current in the 2-Ω resistor, and I_3 be the current in the 10-Ω resistor. Write three equations for I_1, I_2, and I_3 that can be solved for these currents. *Do not* attempt to solve these equations. Be sure to draw a clear diagram in your blue book indicating how you define positive directions.

Answers: Define the directions of positive current flow as shown in the following diagram.

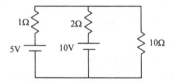

At point a in the circuit, the currents flowing into this point are I_1 and I_2. The current flowing out of this point is I_3. Thus, the point rule gives

$I_1 + I_2 = I_3$

Starting at point a and going clockwise around the left loop, the loop rule gives

$$I_2 \, 2 \, \Omega - 10 \, V + 5 \, V - I_1 \, 1 \, \Omega = 0$$

Starting at point a and going clockwise around the right loop gives

$$-I_3 \, 10 \, \Omega + 10 \, V - I_2 \, 2 \, \Omega = 0$$

These three equations can be solved for I_1, I_2, and I_3.

It is important to first draw the diagram to define the direction of positive current flow in each resistor. The student should realize that this definition is arbitrary at this point and if the wrong direction is chosen, then the current in that resistor will come out to be negative. Once the directions are defined, the point rule and loop rule can be applied to find relationships between the three currents. First, the point rule can be applied at either the branch point at the top of the circuit or the branch point at the bottom, yielding the same result in either case. In applying the point rule, it is important to keep in mind which currents are flowing into the branch point and which currents are flowing out. The loop rule can be applied to the left-hand loop, the right-hand loop, or the outer loop. Any two of these will suffice. In applying the loop rule, it is important to remember that (1) potential is gained in crossing a battery from negative to positive (that is, from the end with the short line to the end with the long line), and (2) potential is lost in crossing a resistor in the direction of the current (that is, the current flows from the more positive side of the resistor to the more negative side). In addition, the student should realize that the three possible loop equations are not independent and so cannot be the three equations that answer the problem. That is, two of the loop equations and the point equation are needed. Because solving these equations can be complicated and time consuming and because I am not interested in testing the algebra skills of the students, I do not require that they solve the equations.

5. A long wire lies along the z axis and carries a current of 1 A in the $+z$ direction.

   a. What is the direction and magnitude of the magnetic field at the point ($x = 0$, $y = 1$ m, $z = 0$)?

   b. A charge of $q = 2C$, which is at the same point as in part a, is moving at a speed of 3 m/s in the $+x$ direction. What is the magnitude and direction of the force on this charge?

   c. A charge of $q = -2$ C (note the negative charge), which is at the same point as in part a, is moving at a speed of 3 m/s in the $+y$ direction. What is the magnitude and direction of the force on this charge?

   d. A charge of $q = 2$ C, which is at the same point as in part a, is moving at a speed of 3 m/s in the $-z$ direction (note the negative direction). What is the magnitude and direction of the force on this charge?

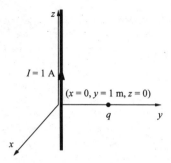

Answers:

a. The magnetic field near a long straight wire carrying current I is

$$B = \frac{\mu_0 I}{2\pi r} = \frac{\left(4\pi \times 10^{-7}\, NA^{-2}\right)1A}{2\pi(1\ \text{m})}$$

This magnetic field forms circular field lines around the wire. The direction is given by the right-hand rule, where the thumb is in the direction of the current and the fingers curl in the direction of the magnetic field. In this case, the magnetic field is in the $-x$ direction at the point $(0, 1\ \text{m}, 0)$.

b. Because the field is in the $-x$ direction and the velocity is in the $+x$ direction, there is no perpendicular component of the magnetic field relative to the velocity. Thus, the force is zero. Another way to see this is to note that the angle between the velocity and the magnetic field is $180°$. Thus, the force is

$$F = qvB \sin\phi = (2C)(3\ \text{m}/\text{s})B\sin 180° = 0$$

c. In this part, the velocity is perpendicular to the field, so that $\sin\theta = \sin(90°) = 1$. Thus, the magnitude of the force is

$$F = qvB = (2C)(3\ \text{m}/\text{s})\frac{\left(4\pi \times 10^{-7}\, NA^{-2}\right)1A}{2\pi(1\ \text{m})}$$

The direction is given by the right-hand rule. Putting your right fingers in the direction of the velocity $(+y)$ and curling them in the direction of the field $(-x)$ puts your thumb in the $+z$ direction. However, the charge is negative, so the force is in the $-z$ direction.

d. As in part c, the velocity is perpendicular to the field, and the magnitude of the force is

$$F = qvB = (2C)(3\ \text{m}/\text{s})\frac{\left(4\pi \times 10^{-7}\, NA^{-2}\right)1A}{2\pi(1\ \text{m})}$$

Putting your right fingers in the direction of the velocity $(-z)$ and curling them in the direction of the field $(-x)$ puts your thumb in the $+y$ direction. Because the charge is positive, this is the direction of the force.

a. In this part, the student should first identify the proper formula for the magnetic field around a long wire. This gives the magnitude of the magnetic field. Next, it should be remembered that magnetic field lines in this case are concentric circles around the wire. At the point chosen, this gives a magnetic field direction that is either in the $+x$ or $-x$ direction. Which direction is correct is given by the right-hand rule.

b. In this part, the student should first ask what is causing a force. The only source present is the magnetic field. The student must then identify the correct formula for the force on a charge moving in a magnetic field, keeping in mind the dependence on the angle between the velocity and the magnetic field. In this part, because the charge is moving exactly opposite to the magnetic field, the angle is 180°, and the force is zero.

c and d. The solutions here are begun the same way as in part b, but now the angle is 90°. Thus, the force is nonzero and must be calculated from the formula. The direction of the force must be perpendicular to both the magnetic field and to the velocity. In both cases (as in general), this gives two possible directions and the right-hand rule is used to distinguish them. The student should realize that the negative charge in part c means that the actual direction of the force is opposite to that given by the right-hand rule.

FINAL EXAM

Keep up with the material; that is, do the readings and homework as assigned and don't fall behind. Secondly, I consider doing homework problems essential in learning physics. A student should make a serious attempt at all homework problems. If he or she has trouble with some of the problems, seek help from fellow students, teaching assistants, or myself. I also distribute solutions to the homework after it has been turned in. I encourage students to compare their solutions with my solutions and spend time understanding the differences.

Be sure to write your name on your blue book. Do all your work in your blue book. Credit will be given for correct answers *only if* the work supports it. No books, papers, notes, or calculators are allowed for this exam. Please do any simple arithmetic (such as $2 \times 3 = 6$) and leave any complex arithmetic as a calculation [such as $(8.2 \times 10^8)(4.39)$]. If anything is unclear about the exam, please ask me. There are 15 problems on three pages of the exam plus two pages of useful information. Each problem is worth 10 points, for a total of 150 points.

1. A guitar string is 1 m long and is stretched to a tension such that the fundamental frequency produced by the string is a middle C ($f = 262$ Hz).

   a. What is the wavelength of the fundamental mode of the string?
   b. What is the wavelength of the sound wave produced? (The velocity of sound waves in air is 344 m/s.)

Answers:

a. The wavelength of the fundamental mode is twice the length of the string; that is, the wavelength is 2 m.

b. The frequency of the sound wave in air is the same as the frequency of the string vibration. Thus, the wavelength of the sound wave is

$$\lambda = \frac{v}{f} = \frac{344 \text{ m/s}}{262 \text{ Hz}}$$

a. This part tests understanding of allowed modes on a string fixed at both ends. The longest-wavelength standing wave (which is called the fundamental) possible in this case is one with a wavelength twice the length of the string; that is, a standing wave that is half one period of a sine wave. In this part, as in many problems, it is important for the student to understand the basic principles that apply and ignore extraneous information (such as the frequency, which is needed for part b).

b. Here, the important basic principle is that the sound waves produced by the guitar string have the same frequency and NOT the same wavelength as the standing waves of the string. Then the frequency and velocity of the sound waves are known, and the wavelength can be calculated easily.

2. A 2-μC charge is at $x = -1$ m and a 1-μC charge is at $x = 1$ m (both are on the x axis).

   a. What are the magnitude and direction of the electric field at the origin ($x = 0$)?

   b. If a -3-μC charge is then placed at the origin, what is the electric force on it?

Answers:

a. Because both charges are positive, the electric field due to each charge points radially outward from the charge. Thus, the field due to the 1-μC charge points in the $-x$ direction and has a magnitude of

$$E_1 = \frac{Q_1}{4\pi\varepsilon_o r^2} = \frac{\left(9 \times 10^9 \text{ Nm}^2\text{C}^{-2}\right)1 \text{ μC}}{(1 \text{ m})^2} = 9 \times 10^3 \text{ N/C}$$

The field due to the 2-μC charge points in the $+x$ direction and has a magnitude of

$$E_2 = \frac{Q_2}{4\pi\varepsilon_o r^2} = \frac{\left(9 \times 10^9 \text{ Nm}^2C^{-2}\right)2 \text{ μC}}{(1 \text{ m})^2} = 18 \times 10^3 \text{ N/C}$$

Because electric field is a vector, the total electric field in the $+x$ direction is

$$E = E_2 - E_1 = 18 \times 10^3 \text{ N/C} - 9 \times 10^3 \text{ N/C} = 9 \times 10^3 \text{ N/C}$$

b. The magnitude of the force on a charge q in an electric field E is $F = qE$, giving in this case, $F = (3 \text{ μC})(9 \times 10^3 \text{ N/C}) = 0.027$ N. Because the charge is negative and the electric field is in the $+x$ direction, the force is in the $-x$ direction.

a. This part asks for the electric field from two charges. The student must first determine the correct formula for calculating the electric field due to one point charge. Then the student should remember the superposition principle; that is, the electric field due to two charges is the sum of the electric field due to each charge. Finally, it is necessary to remember that the electric field is a vector quantity, and the two contributions must be added appropriately, which in this case means taking proper account of the sign of the x component of the field.

b. Here the student should recognize that the electric force is due to the electric field and is given by $F = qE$. In addition, because force is a vector quantity, it is necessary to specify both the magnitude and direction of the force, keeping in mind that the negative charge implies that the force is opposite to the electric field.

3. A proton starts from rest and is accelerated through an electric potential of 100 V.

   a. How much work is done on the proton by the electric field?

   b. What is the final velocity of the proton?

Answers:

a. Electric potential is work per unit charge, so that $U = qV$, where q in this problem is the charge of the proton $(1.6 \times 10^{-19}\,\text{C})$, giving $U = (1.6 \times 10^{-19}\,\text{C})\,100\,\text{V} = 1.6 \times 10^{-17}\,\text{J}$. Because the magnitude of the charge of the proton is the same as that of the electron, this can also be expressed as 100 eV.

b. By conservation of energy, the initial energy is equal to the final energy. The initial energy is the potential energy of the proton, and the final energy is entirely kinetic energy, giving

$$\tfrac{1}{2}mv^2 = U = qV \Rightarrow v = \sqrt{\frac{2qV}{m}} = \sqrt{\frac{2(1.6 \times 10^{-17}\,\text{J})}{1.67 \times 10^{-27}\,\text{kg}}}$$

a. For this part, the student should remember that electric potential and electric potential energy are closely related concepts; namely, electric potential is electric potential energy per unit charge. Thus, the change in potential energy is the charge of the proton (which is given on the exam) times the potential difference. The change in potential energy is actually work done on the proton by the electric force.

b. Because the functional form of the electric potential is not given, the form of the electric force cannot be determined. In particular, the acceleration is not necessarily constant, so constant acceleration methods cannot be applied. The student should realize that energy methods are appropriate because static electric forces are conservative. By the work energy theorem, the work done (calculated in part b) is equal to the change in kinetic energy. It is possible to assume that the electric field is constant over some length, deduce the force on the proton and hence its acceleration (the mass of the proton is given on the exam), and apply constant acceleration equations. Because the result is independent of the functional form of the electric force and depends only on the potential difference, the constant acceleration method will yield the correct answer. However, this method is based on an assumption that is not necessarily correct, and this method is more laborious than the energy method, and thus is not a good solution.

4. Two 10-μF capacitors are connected in series and then connected to a 10-V battery.

a. What is the charge on each capacitor?

b. What is the total stored energy in the two capacitors?

The same two capacitors are now connected in parallel and then reconnected to the same 10-V battery.

c. What is the charge on each capacitor now?

d. What is the total stored energy in the two capacitors?

Answers:

a. When the capacitors are in series, the charge is the same on each of them and is equal to the charge on the equivalent capacitance of 5 μF, giving

$$Q = CV = (5 \mu F)10 \text{ V} = 50 \mu C$$

b. The total stored energy is the sum of the stored energy on each capacitor. There are several expressions for the stored energy, but because in this case we know the capacitance and charge of each capacitor, the easiest solution is

$$U = U_1 + U_2 = \frac{Q_1^2}{2C_1} + \frac{Q_2^2}{2C_2} = \frac{(50 \mu C)^2}{2(10 \mu F)} + \frac{(50 \mu C)^2}{2(10 \mu F)} = 250 \mu J$$

c. When the capacitors are in parallel, they have the same potential across them, namely, 10 V. The charge on each capacitor is thus given by

$$Q = CV = (10 \mu F)10 \text{ V} = 100 \mu C$$

d. The total stored energy is the sum of the stored energy on each capacitor. Any of the several expressions will work here, including

$$U = U_1 + U_2 = \tfrac{1}{2}Q_1V_1 + \tfrac{1}{2}Q_2V_2 = \tfrac{1}{2}(100 \mu C)10 \text{ V} + \tfrac{1}{2}(100 \mu C)10 \text{ V} = 1000 \mu J$$

a. First, the student should remember that a series connection means that the two capacitors are connected on one end, not connected to each other on the other end, and nothing else is connected to the common end. Because the common ends are tied together and because charge is conserved and has nowhere else to go, the charge on the plate of one capacitor is the negative of the charge on the connected plate of the other capacitor, so that the magnitude of the charge on each capacitor is the same. There are then several ways to attack this problem.

First, the equivalent capacitance could be calculated and the charge on the equivalent capacitance determined. By the definition of what is meant by the equivalent capacitance, the charge that must have flowed into the equivalent capacitance from the battery is the same as the charge that flowed into the actual capacitors. Because the actual capacitors are in series, this charge must have flowed onto one plate of one of them and off the opposite plate of the second one. Thus, this is the charge on each capacitor. A second method that works in this problem is to realize that because each capacitor has the same charge and the same capacitance, the voltage across each capacitor is the same, namely, 5 V (half of the voltage supplied by the battery). The charge of each capacitor is then calculated from $Q = CV$. This second method works for the problem given, whereas the first method works for any values of the two capacitors.

b. The total stored energy is the sum of the stored energies of the capacitors. There are several formulas for the stored energy of a capacitor that depend on various combinations of Q, C, and V. It is best to choose one that depends on information that is already known, which in this case is Q (from part a) and C (given in the problem).

c. First, the student must remember that the definition of a parallel connection is that both sides of the capacitors are connected together. These common connections are then connected to each side of the battery. Next, the student should realize that because electric potential is path independent, the potential across each capacitor is the same as the potential across the battery, that is, 10 V. Thus, the voltage across each capacitor is known (10 V) and the capacitance is known, so that the charge can be calculated from $Q = CV$, which is always true for a capacitor.

d. Again, there are several formulas for the stored energy. In this case, any of them will work because we already know Q, C, and V for each capacitor.

5. A 1.5-V battery with an internal resistance of 0.2 Ω is connected in parallel to a 1.6-V battery with an internal resistance of 0.1 Ω. These two batteries are then connected to a light bulb with a resistance of 1 Ω.

   a. Draw a circuit diagram and label the current in each resistor (including the lightbulb). Give a name (e.g., I_1) and direction for each current.

   b. Write down equations that can be solved for these currents (you do NOT need to solve these equations).

   c. In terms of the currents in the resistors and the other variables in the problem, what is the power dissipated in the lightbulb?

Answers:

a.

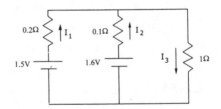

b. From the point rule, we have

$$I_1 + I_2 - I_3 = 0$$

From the loop rule around the left loop, we have

$$1.5\,\text{V} - I_1(0.2\,\Omega) - I_3(1\,\Omega) = 0$$

From the loop rule around the right loop, we have

$$1.6\,\text{V} - I_2(0.1\,\Omega) - I_3(1\,\Omega) = 0$$

These three equations may be solved for the three currents. The equation from applying the loop rule around the outer loop can be substituted for either of the other two loop equations.

c. The power dissipated in a resistor is $I_2 R$, giving in this case

$$P = I^2 R = I_3^2 (1 \ \Omega)$$

a. It is important for the student to first translate the description of the circuit into a circuit diagram. First, the student should remember that a real battery (that is, one with internal resistance) is modeled as an ideal battery in series with a resistor. In this problem, the two real batteries are then connected in parallel, meaning that both ends of the two batteries are connected together. Next, this parallel connection is connected to the 1-W resistor that represents the lightbulb. Finally, the student must clearly define and label the current in each resistor. This is important because the equations developed in part b depend exactly on how the currents are defined. The student should realize that the direction defined for any given current is arbitrary, in that if the current is actually flowing in the opposite direction then its value will come out negative when the equations are solved.

b. Once the directions are defined, the point rule and loop rule can be applied to find relationships between the three currents. First, the point rule can be applied at either the branch point at the top of the circuit or the branch point at the bottom, yielding the same result in either case. In applying the point rule, it is important to keep in mind which currents are flowing into the branch point and which currents are flowing out. The loop rule can be applied to the left-hand loop, the right-hand loop, or the outer loop. Any two of these will suffice. In applying the loop rule, it is important to remember that (1) potential is gained in crossing a battery from negative to positive (that is, from the end with the short line to the end with the long line), and (2) potential is lost in crossing a resistor in the direction of the current (that is, the current flows from the more positive side of the resistor to the more negative side). In addition, the student should realize that the three possible loop equations are not independent and so cannot be the three equations that answer the problem. That is, two of the loop equations and the point equation are needed. Because solving these equations can be complicated and time consuming and because I am not interested in testing the algebra skills of the students, I do not require that they solve the equations.

c. The power dissipated in a resistor can be written in several forms, which depend on various combinations of I, R, and V. In this case, it is best to choose the one that has I and R ($P = I_2 R$), because we already know what I is (in principle from part b) and what R is (from the problem).

6. An electron is in a uniform electric field of 10 V/m that is pointing upwards ($+z$ direction). It has a velocity of 1×10^4 m/s in the $+x$ direction. What is the direction and magnitude of the uniform magnetic field that must be applied so that the net force on the electron is zero (neglect gravity)?

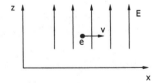

Answer: Because the charge of the electron is negative, the force due the electric field is opposite to the electric field, that is, down. To make the net force zero, the force due the magnetic force must be up. Because the magnetic force is perpendicular to both the velocity of a charge and to the magnetic field, the magnetic field must be either into or out of the paper. By the right-hand rule (keeping in mind that the charge of the electron is negative), the magnetic field must be out of the paper, that is, in the $-y$ direction. The magnitude of the magnetic field is found by equating the electric and magnetic forces, giving

$$F_E = F_B \Rightarrow qE = qvB \Rightarrow B = \frac{E}{v} = \frac{10 \text{ V/m}}{10^4 \text{ m/s}} = 0.001 \text{ T}$$

First, the student should identify the forces in the problem. Because there is an electric field and a charge, there is an electric force given by $F_E = qE$, where q is the charge of the electron (which is given on the exam). The direction of the force is opposite to the electric field because the charge of the electron is negative. There is also a magnetic force because the charge is moving in a magnetic field. The magnetic force is given by $F_B = qvB \sin(\phi)$, where ϕ is the angle between the velocity and the magnetic field. Because the problem says to neglect gravity, there are no other forces. Because the electric force is not zero, the only way for the total force to be zero is for magnetic force to be equal in magnitude and opposite in direction from the electric force (the vector nature of force is important here). Thus, the magnetic force should be up (in the $+z$ direction). The magnetic force is perpendicular to both the magnetic field and the velocity, so it is either into the page or out of the page. Using the right-hand rule and keeping in mind that the direction must be reversed due to the negative charge of the electron, it is found that the magnetic field must be out of the page. Actually this only specifies the component of the magnetic field perpendicular to the velocity, because any component parallel to the velocity does not contribute to the force. Thus, the magnetic field could be given any component parallel to the velocity (that is, in the x direction), but by far the simplest choice is to make the x component zero. With the x component zero, the magnetic field is perpendicular to the velocity ($f = 90°$). Equating the magnitudes of the electric and magnetic fields gives $qE = qvB$. The charge of the electron cancels out of the equation, which means doing the problem algebraically is simpler than it doing arithmetically (that is, actually calculating a value for the electric force and then setting the magnetic force equal to this value).

7. Two long, parallel wires are 2 m apart. The top wire has a current of 1 A to the left. The bottom wire has a current of 2 A to the left. What is the magnitude and direction of the magnetic field at a point halfway between the wires (see diagram)?

Answer: The magnetic field at a distance r from a long, straight wire is

$$F = \frac{\mu_0 I}{2\pi r}$$

The field lines form circles centered on the wire with a direction given by the right-hand rule. Thus, at the point of interest, the magnetic field from the top wire is out of the page, and the magnetic field from the bottom wire is into the page. Because the field from the bottom wire is greater (the current is larger and the distance the same), the total magnetic field is into the page and is equal to

$$B = B_{\text{bottom}} - B_{\text{top}} = \frac{4\pi \times 10^{-7}\, NA^{-2}(2\text{ A})}{2\pi(1\text{ m})} - \frac{4\pi \times 10^{-7}\, NA^{-2}(1\text{ A})}{2\pi(1\text{ m})} = 2 \times 10^{-7}\text{ T}$$

First, the student should realize that the superposition principle applies; that is, the magnetic field is the sum of the fields due to each of the wires, keeping in mind that the magnetic field is a vector and must be added as such. The student should then identify the field due to a long straight wire from among the formulas given on the exam. Next, because magnetic field is a vector, the direction is important. Magnetic field lines around a long wire are concentric circles with the sense of the magnetic field given by the right-hand rule. Thus, the field from the top wire points out of the page and the field from the bottom wire points into the page. Then, all that is left is to calculate the magnitude of the field from each wire at the desired point and combine them with the appropriate sign.

8. A circular loop of wire has an area of 10 cm² and a total resistance of 10 Ω. The loop is in a magnetic field of 1 Tesla (1 T) that is normal to the area of the loop. The field is decreasing at the rate of 0.1 T/s. What is the magnitude and direction of the current in the loop? You need to make a careful drawing to indicate the direction. You also need to give the reason you chose the direction of current you did.

Answer: This problem is an example of Faraday's law of induction. The changing flux through the loop induces an emf E around the loop given by

$$E = \frac{\Delta\Phi}{\Delta t} = \frac{A\Delta B}{\Delta t} = 10\text{ cm}^2 \left(\frac{1\text{ m}}{100\text{ cm}}\right)^2 0.1\text{ T/s} = 10^{-4}\text{ V}$$

By Ohm's law, the current in the loop is

$$I = \frac{E}{R} = \frac{10^{-4}\text{ V}}{10\text{ Ω}} = 10^{-5}\text{ A}$$

Looking down on the loop with the field into the page gives

By Lenz's law, the induced current opposes the change in flux. Because the field is decreasing, if the field is into the page, the change is out of the page, and the current must be clockwise to oppose the change in flux.

The two ways to cause a current to flow in a circuit are either to supply a potential difference (such as a battery) or to have a changing magnetic flux through the circuit. Because there is no battery in this problem, the current must be due to the changing flux. The emf around a loop due to a changing flux is given by Faraday's law of induction. This is true no matter what the cause of the changing flux. In this problem, the flux is changing because the magnetic field is changing. Using Ohm's law to equate the induced emf (E) to the current in the loop ($E = IR$) gives the desired current. The direction of the current is given by Lenz's law. In applying Lenz's law, it is first important to realize that the current induced in the loop also produces a magnetic field (different from the original field). Lenz's law says that this induced field is opposite to the change of the original field. Note that it is the change in the original field that is important, not the direction of the original field. Also, the student should recognize that this problem has extraneous information given, namely, the magnitude of the original field. Because the magnetic flux is changing due to the change of the magnetic field, what counts is the rate of change of the magnetic field, and not its magnitude.

9. Green light has a wavelength of 500 nm in air.

   a. What is the frequency of this light?
   b. What is the wavelength of this light in water ($n = 1.33$)?

Answers:

a. The velocity of light waves in air is $c = 3 \times 10^8$ m/s. The frequency is related to the velocity and wavelength by

$$f = \frac{c}{\lambda} = \frac{3 \times 10^8 \, \text{m/s}}{500 \times 10^{-9} \, \text{m}} = 6 \times 10^{14} \, \text{Hz}$$

b. In water, the frequency of the light is the same, and the velocity is $v = \frac{c}{n}$. Thus, the wavelength in water is

$$\lambda = \frac{v}{f} = \frac{c}{nf} = \frac{3 \times 10^8 \, \text{m/x}}{1.33(6 \times 10^{14} \, \text{Hz})} = 376 \, \text{nm}$$

a. Given the wavelength, the student should realize that the velocity of the wave is necessary to find the frequency. Then the student should realize that the velocity of light in air is very close to its velocity in vacuum, namely 3×10^8 m/s.

b. For this part, the student needs to understand two things. First, the velocity of light in water is smaller than it is in vacuum; that is, $v = c/n$. Second, the light in the water has the same frequency (NOT wavelength) as the light in air.

10. There is a planar interface between a sheet of glass with an index of refraction of 1.5 and a pool of water with an index of refraction of 1.33. A light ray coming from the glass is incident on this boundary at an angle of 10° to the normal.

   a. What is the angle of reflection?
   b. What is the angle of the light ray that passes through to the water?
   c. What is the maximum angle of incidence that the light ray can have and still have some light pass through to the water?

Answers:

a. The angle of reflection is equal to the angle of incidence, that is, 10°.

b. The angle of refraction is given by Snell's law:

$$\sin \phi_b = \frac{n_a}{n_b} \sin \phi_a = \frac{1.5}{1.33} \sin(10°)$$

c. The critical angle is given by the criterion that the refracted angle be 90°, giving

$$\sin \phi_{crit} = \frac{n_b}{n_a} = \frac{1.33}{1.5}$$

a. This is a straightforward application of the law of reflection.

b. The student should recognize this as refraction of light, which is governed by Snell's law.

c. The student should remember that when light passes from a higher index of refraction to a lower index of refraction, there is a range of incident angles for which Snell's law cannot be satisfied (it gives the sine of the refracted angle as greater than 1, which is impossible). This does not mean that Snell's law is wrong—it just means that Snell's law doesn't apply, because there is no refracted ray; that is, the light is totally internally reflected. The critical angle is the incident angle, such that Snell's law gives the sine of the angle of refraction as its maximum value, namely 1.

11. A thin lens has a focal length of 20 cm. An object that is 2 m tall is placed on the axis of the lens at a position 10 cm to the left of the lens.

   a. Draw a principle-ray diagram showing where the image is formed.
   b. Calculate the position of the image.
   c. Is the image real or virtual?
   d. What is the size of the image?

Answers:

a. (Note that the vertical and horizontal scales are different for clarity.)

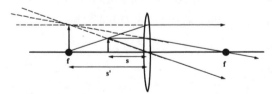

b. Because the object is on the same side as the light entering the lens, s is positive ($s = 2$ cm). Because this is a converging lens, f is positive ($f = 20$ cm). By the lens formula,

$$\frac{1}{s} + \frac{1}{s'} = \frac{1}{f} \Rightarrow \frac{1}{s'} = \frac{1}{f} - \frac{1}{s} = \frac{1}{20 \text{ cm}} - \frac{1}{10 \text{ cm}} = -\frac{1}{20 \text{ cm}} \Rightarrow s' = -20 \text{ cm}$$

Because the image distance is negative, the image is on the opposite side of the lens as the outgoing light, that is, on the left side as drawn.

c. Because no actual light rays pass through the image position, the image is virtual. This is confirmed by the negative image distance.

d. The image size is given by

$$m = \frac{y'}{y} = -\frac{s'}{s} \Rightarrow y' = -\frac{s'}{s}y = -\frac{-20 \text{ cm}}{10 \text{ cm}}2 \text{ cm} = 4 \text{ cm}$$

a. The drawing of the diagram is important in this case to understand where the image is and what it looks like. Doing this carefully will greatly help in checking that the numerical parts of this problem are done correctly. Only two of the principle rays are needed to determine the size and location of the image. However, drawing more principle rays serves as a consistency check.

b, c, and d. The most important thing about the numerical solution of this problem is understanding the sign conventions of the various quantities. If the student uses an incorrect sign (for example, making the focal length negative in part b), then she or he will get an answer that is inconsistent with the diagram that she or he drew in part a. If the student uses the correct sign, she or he will get an image position of -20 cm, which means that the image is 20 cm from the lens on the side opposite to the outgoing light, that is, the image should be at the left focus. If there is a slight disagreement, this may be due to small errors in drawing the diagram. Large disagreements indicate either incorrect answers or significant errors in the diagram. Thus, checking that answers are consistent with the diagram is important.

12. A spaceship flies past Mars with a speed of 0.964c relative to the surface of Mars. When the spaceship is directly overhead at an altitude of 1500 km, a very bright signal light on the Martian surface blinks on and then off. An observer on Mars measures that the signal light was on for 8.00×10^{-5} s. What is the duration of the light pulse measured by the pilot of the spaceship?

Answer: To the observer on the spaceship, Mars is moving with $u = 0.964c$, and thus clocks on Mars appear to run slow by a factor of γ. Thus the time between events is increased by this factor and the time measured by the pilot is

$$\Delta t = \gamma \Delta t' = \frac{1}{\sqrt{1 - \dfrac{u^2}{c^2}}}t' = \frac{1}{\sqrt{1 - 0.964^2}}\left(8 \times 10^{-5} \text{s}\right) = 3.01 \times 10^{-4}\text{s}$$

Because the problem is about duration of the signal light, it is about measuring a time. Thus, relativistic time dilation is the relevant physics. The important point about time dilation is that an observer measures moving clocks to run slower than stationary clocks by a factor of

$$\gamma = \frac{1}{\sqrt{1 - \dfrac{u^2}{c^2}}}$$

Because moving clocks run slower, the duration of moving events is longer by the same factor of γ. The student must recognize that to the pilot, the clocks on his spaceship are stationary and those on the planet are moving. The student also needs to recognize that the altitude is extraneous information not needed to solve the problem.

13. The energy levels of hydrogen are given by $E_n = -13.6\ \text{eV}/n^2$. The first line of the Balmer series is a transition from $n = 3$ to $n = 2$. What is the frequency of the light that is emitted in this transition?

Answer: The energy of the photon emitted in this transition is equal to the difference in energy between the initial and final levels; that is,

$$E_\gamma = E_i - E_f = E_3 - E_2 = \frac{-13.6\ \text{eV}}{3^2} - \frac{-13.6\ \text{eV}}{2^2} = \left(\frac{1}{4} - \frac{1}{9}\right)13.6\ \text{eV} = \frac{5}{36}13.6\ \text{eV}$$

The frequency of light that this corresponds to is given by

$$E_\gamma = hf \Rightarrow f = \frac{E_\gamma}{h} = \frac{\frac{5}{36}13.6\ \text{eV} \times \frac{1\ \text{eV}}{1.6\times10^{-19}\ \text{J}}}{6.63\times10^{-34}\ \text{Js}}$$

The student should realize that two basic physics principles apply here. First, the energy of a photon emitted in an atomic transition is equal to the difference in energy of the two energy levels. Part of this principle is that light is emitted in quanta, that is, as discrete particles with a definite energy. The second principle is the relationship between the energy of a photon and the frequency of the light; that is, $E = hf$.

14. What is the atomic structure (that is, which energy levels are occupied by electrons) of the ground state of phosphorus ($Z = 15$)?

Answer: Because there are two electrons per energy level (one with spin up and one with spin down) and because there is one $1s$ level, one $2s$ level, and three $2p$ levels, one $3s$ level, and three $3p$ levels, the ground state of phosphorus is $1s^2 2s^2 2p^6 3s^2 3p^3$.

The student first needs to understand the labeling of the energy levels of an atom. Next, the student needs to understand that the Pauli exclusion principle allows only two electrons (one with spin up and one with spin down) per energy level.

15. The isotope ^{238}U of uranium ($Z = 92$) decays by alpha emission to an isotope of thorium.

    a. What are Z, N, and A for this isotope of thorium?
    b. The mass of ^{238}U is 238.0508 u and the mass of an alpha particle is 4.0026 u. If the energy released in this decay is 4.268 MeV, what is the mass of the resultant thorium isotope (in atomic mass units—note that $1\ \text{u} = 1.66056 \times 10^{-27}\ \text{kg}$).

Answers:

a. An alpha particle is a helium nucleus consisting of two protons and two neutrons. Thus, emitting an alpha particle lowers Z (the number of protons) by 2, N (the number of neutrons) by 2, and A (the number of protons plus neutrons) by 4. Thus, $Z = 92 - 2 = 90$, $N = 146 - 2 = 144$ (^{238}U has $N = 238 - 92 = 146$), and $A = 238 - 4 = 234$.

b. By conservation of energy, the energy released is equal to the difference between the rest energies of the uranium and alpha particle and the rest energy of the thorium. Thus,

$$M_U c^2 - M_\alpha c^2 - M_{Th} c^2 = E$$

$$\Rightarrow M_{Th} = M_U - M_\alpha - \frac{E}{c^2}$$

$$= 238.0508 \text{ u} - 4.0026 \text{ u} - \frac{4.268 \times 10^6 \text{ eV}}{\left(3 \times 10^8 \text{ m/s}\right)^2} \frac{1.6 \times 10^{-19} \text{ J}}{\text{eV}} \frac{1 \text{ u}}{1.66056 \times 10^{-27} \text{ kg}}$$

---

a. In this problem, the student needs to keep straight the concepts of atomic number Z (the number of protons in the nucleus), neutron number N (the number of neutrons), and atomic mass number A (the sum of the number of protons and neutrons). Thus, uranium has 92 protons ($Z = 92$), 144 neutrons (238 – 92), and 238 protons and neutrons. Second, the student needs to know that an alpha particle has two protons and two neutrons. Thus, emission of an alpha particle reduces Z and N by 2 each (and, thus, A by 4).

b. This part of the problem is based on the equivalence of mass and energy. Energy is conserved, so the energy before the decay (that is, the rest energy of the uranium) is equal to the energy after the decay (the rest energies of the alpha particle and thorium plus the energy released as kinetic energy of the final state products). The rest energy of a mass M is $E = Mc^2$.

Useful Information

The following information appears with the midterm and the final exam

$$v = \lambda f$$

$$v = \sqrt{\frac{T}{\mu}}$$

$$v = \sqrt{\frac{B}{\rho}}$$

$$v = \sqrt{\frac{Y}{\rho}}$$

$$y(x,t) = A \sin\left[2\pi\left(\frac{t}{T} - \frac{x}{\lambda}\right)\right]$$

$$f_n = \frac{nv}{2L} \quad (n = 1,\ 2,\ 3,\ ...)$$

$$f_n = \frac{nv}{4L} \quad (n = 1,\ 3,\ 5,\ ...)$$

$$p_{max} = BkA$$

$$I = \frac{p_{max}^2}{2\rho v}$$

$$\beta = 10\log\frac{I}{I_0}$$

$$I_0 = 10^{-12}\,W/m^2$$

$$f_{beat} = \left|f_1 - f_2\right|$$

$$m_e = 9.11 \times 10^{-31}\,\text{kg}$$

$$m_p = 1.67 \times 10^{-27}\,\text{kg}$$

$$q_e = -1.6 \times 10^{-19}\,\text{C}$$

$$q_p = 1.6 \times 10^{-19}\,\text{C}$$

$$F = \frac{q_1 q_2}{4\pi\varepsilon_0 r^2}$$

$$\frac{1}{4\pi\varepsilon_0} = 9 \times 10^9\,\text{Nm}^2\text{C}^{-2}$$

$$\varepsilon_0 = 8.85 \times 10^{-12}\,\text{N}^{-1}\text{C}^2\text{m}^{-2}$$

$$\vec{E} = \frac{\vec{F}}{q}$$

$$E = \frac{q}{4\pi\varepsilon_0 r^2}$$

$$\Psi = \sum E_\perp \Delta A = \frac{Q}{\varepsilon_0}$$

$$U = \frac{qq'}{4\pi\varepsilon_0 r}$$

$$V = \frac{U}{q'}$$

$$V = \frac{1}{4\pi\varepsilon_0} \sum \frac{q_i}{r_i}$$

$$E = \frac{\Delta V}{\Delta s}$$

$$Q = CV$$

$$C = \frac{\varepsilon_0 A}{d}$$

$$C_{parallel} = C_1 + C_2$$

$$\frac{1}{C_{series}} = \frac{1}{C_1} + \frac{1}{C_2}$$

$$U = \tfrac{1}{2}QV = \tfrac{1}{2}CV^2 = \frac{Q^2}{2C}$$

$$u = \tfrac{1}{2}\varepsilon_0 E^2$$

$$I = \frac{\Delta Q}{\Delta t}$$

$$J = nqv$$

$$\rho = \frac{E}{J}$$

$$R = \frac{\rho \ell}{A}$$

$$V = IR$$

$$P = IV = I^2 R = \frac{V^2}{R}$$

$$\sum_{point} I = 0$$

$$\sum_{loop} V = 0$$

$$R_{series} = R_1 + R_2$$

$$\frac{1}{R_{parallel}} = \frac{1}{R_1} + \frac{1}{R_2}$$

$$F = qvB \sin\phi$$

$$\Phi = \sum B_\perp \Delta A$$

$$R = \frac{mv}{qB}$$

$$F = I\ell B$$

$$\mu_0 = 4\pi \times 10^{-7}\,\text{NA}^{-2}$$

$$B_{wire} = \frac{\mu_0 I}{2\pi r}$$

$$B_{loop} = \frac{\mu_0 NI}{2R}$$

$$B_{solenoid} = \mu_0 nI$$

$$B_{toroid} = \frac{\mu_0 NI}{2\pi r}$$

$$B = \frac{\mu_0}{4\pi} \frac{qv \sin\theta}{r^2}$$

$$\Delta B = \frac{\mu_0}{4\pi} \frac{I\Delta\ell \sin\theta}{r^2}$$

$$\sum B_{\parallel} \Delta s = \mu_0 I_{enclosed}$$

$$= vB\ell$$

$$= -\frac{\Delta\Phi}{\Delta t}$$

$$\sum E_\parallel \Delta s = -\frac{\Delta\Phi}{\Delta t}$$

$$\sum B_\perp \Delta A = 0$$

$$\sum B_\parallel \Delta s = \mu_0\left(I_c + \varepsilon_0 \frac{\Delta\Psi}{\Delta t}\right)$$

$$c = \frac{1}{\sqrt{\varepsilon_0 \mu_0}}$$

$$\frac{c}{v} = n$$

$$n_a \sin\phi_a = n_b \sin\phi_b$$

$$\sin\phi_{\text{critical}} = \frac{n_b}{n_a}$$

$$m = \frac{y'}{y} = -\frac{s'}{s}$$

$$\frac{1}{s} + \frac{1}{s'} = \frac{1}{f}$$

$$f = \frac{R}{2}$$

$$\frac{n}{s} + \frac{n'}{s'} = \frac{n' - n}{R}$$

$$m = \frac{y'}{y} = -\frac{ns'}{n's}$$

$$M_{\text{microscope}} = m_1 M_2 = \frac{(25cm)s_1'}{f_1 f_2}$$

$$M_{\text{telescope}} = \frac{u'}{u} = -\frac{f_1}{f_2}$$

$$\Delta t = \frac{\Delta t'}{\sqrt{1 - u^2/c^2}}$$

$$\ell = \ell'\sqrt{1 - u^2/c^2}$$

$$x' = \frac{x - ut}{\sqrt{1 - u^2/c^2}}$$

$$y' = y; \quad z' = z$$

$$t' = \frac{t - ux/c^2}{\sqrt{1 - u^2/c^2}}$$

$$v' = \frac{v - u}{1 - uv/c^2}$$

$$v = \frac{v' + u}{1 + uv'/c^2}$$

$$\vec{p} = \frac{m\vec{v}}{\sqrt{1 - v^2/c^2}}$$

$$\vec{F} = \frac{\Delta\vec{p}}{\Delta t}$$

$$E = K + mc^2 = \frac{mc^2}{\sqrt{1 - v^2/c^2}}$$

$$E = hf$$

$$h = 6.63 \times 10^{-34}\ Js$$

$$p = \frac{h}{\lambda}$$

$$\Delta x \Delta p_x \geq \frac{h}{2\pi}$$

$$\frac{\Delta N}{\Delta t} = -\lambda N$$

$$t_{\frac{1}{2}} = \frac{0.693}{\lambda}$$

If an object is on the same side of a lens or mirror as the incoming light, s is positive; otherwise it is negative.

If an image is on the same side of a lens or mirror as the outgoing light, s' is positive; otherwise, it is negative.

For a spherical mirror, the focal length is positive if the center of curvature is on the same side as the outgoing light, otherwise, it is negative.

For a converging lens, the focal length is positive; for a diverging lens, it is negative.

CHAPTER 19

CALIFORNIA STATE UNIVERSITY, FRESNO

PHYSICS 4A: MECHANICS AND WAVE MOTION

Lauren J. Novatne, Teaching Assistant

T HERE ARE TWO MIDTERMS AND A FINAL IN THIS COURSE. AN EXAMPLE OF ONE midterm and the final are given.

The primary objectives of the course are to prepare students for the core engineering and science courses, and to develop a basic understanding of mechanics and wave motion.

Key competencies to be acquired are the ability to understand conceptually the nature of motion, work, and energy, and to calculate quantities of measure for each of them. The quality of thought should include, but not be limited to, the analytic ability to assess the needed quantity of measure to solve a problem, as well as the synthetic ability to create the problem-solving strategy.

Because the exams ask both conceptual questions and require computation of physical quantities related to the topics, it is hoped that they indicate how each of the course objectives have been accomplished.

I would encourage students to strive toward understanding the concepts first, then to focus on accurate computation as a secondary emphasis.

MIDTERM EXAM

I try to design my exams so that the best students will be able to complete all items and 80 percent of the students will be able to complete 80 percent of the material.

1. If you drop an object in the absence of air resistance, it accelerates downward at 9.8

m/s². If, instead, you throw the object downward, its downward acceleration after release is:

a. less than 9.8 m/s².

b. 9.8 m/s².

c. greater than 9.8 m/s².

Answer: b

The acceleration is constant, independent of initial velocity.

2. An object moves from one spot to another. After the object gets to its destination, the displacement is:

a. either greater than or equal to

b. always greater than

c. always equal to

d. either smaller than or equal to

e. always smaller than

f. either smaller or larger than

the distance it traveled.

Answer: d

The displacement is the vector pointing from the initial to the final position. The path the object traveled could have been longer than the displacement from the initial position.

3. A particle has a position given by the equation $x = 5 - 34t + 5t^4$. The units of the coefficients are m, m/s, and m/s², respectively. Find the velocity and acceleration of the particle.

Answers:

Velocity: $34 + 20t^3$

Acceleration: $60t^2$

To begin, first differentiate $x(t)$ once with respect to t. This will give the velocity of the particle: $34 + 20t^3$.

Next, to get the acceleration, differentiate $x(t)$ with respect to t again. This gives $60t^2$.

4. A plane is flying at a constant elevation of 1200 m with a constant speed of 430 km/h. At what angle ϕ should the cargo mate drop a package intended to land on the ground at a specific point?

Answer:

$$\phi = \tan^{-1}\left(\frac{x}{y}\right) = \frac{1869}{1200}$$

The correct equation to begin with is:

$$y - y_0 = (v_0\sin\theta_0)t - \tfrac{1}{2}gt^2$$

where $\theta_0 = 0$.

Solving for t will yield 15.65 s.

Next, the solution requires that the horizontal distance be calculated:

$$x - x_0 = (v_0\cos\theta)t = 1869 \text{ m. If } x_0 = 0, \text{ then } x = 1869 \text{ m.}$$

The angle of sight is then:

$$\phi = \tan^{-1}\left(\frac{x}{y}\right) = \frac{1869}{1200}$$

5. A person of mass 72.2 kg stands on a platform scale in an elevator that is going up. What will the scale read at the times of

a. upward acceleration of magnitude 3.20 m/s^2?

b. constant speed upward?

c. downward acceleration of magnitude 3.20 m/s^2?

Draw free-body diagrams to assist you in solving this problem if needed.

Answers:

a. 920 N

b. 708 N

c. 477 N

a. The correct equation for this portion of the elevator trip is:
$$N = m(g + a) = (72.2 \text{ kg}) (9.80 \text{ m/s}^2 + 3.2 \text{ m/s}^2) = 920 \text{ N}$$

b. The correct equation for this portion of the elevator trip is:
$$N = m(g + a) = (72.2 \text{ kg}) (9.80 \text{ m/s}^2 + 0 \text{ m/s}^2) = 708 \text{ N}$$

c. The correct equation for this portion of the elevator trip is:
$$N = m(g + a) = (72.2 \text{ kg}) (9.80 \text{ m/s}^2 - 3.2 \text{ m/s}^2) = 477 \text{ N}$$

FINAL EXAM

Students are well advised to understand key concepts first, and then to practice the application of these concepts with problem-solving skills. Good study habits include the repetition of exposure to material. The number of ways to be exposed (reading, lecture, homework, and lab activities) is as important as the number of times the student is exposed to the topics (reading the chapter two or three times).

1. An object moves from one spot to another. After the object gets to its destination,

the displacement is:

a. either greater than or equal to

b. always greater than

c. always equal to

d. either smaller than or equal to

e. always smaller than

f. either smaller or larger than

the distance it traveled.

Answer: d

> The displacement is the vector pointing from the initial to final position. The path the object traveled could have been longer than the displacement from the initial position.

2. By shaking the end of a stretched string, a single pulse is generated. The traveling pulse carries:

a. energy.

b. momentum.

c. energy and momentum.

d. neither of the two.

Answer: c

> A traveling wave carries no mass. By shaking one end, one can get the other end to move. The resulting motion has momentum and energy associated with it, so these quantities are transported through the string.

3. How much work is done by a force $F = (3x \text{ N})I + (4 \text{ N})J$ (with x in meters) that acts on a particle as it moves from coordinates $(2m, 3m)$ to $(3m, 0m)$?

Answer: $W = 45 \text{ J}$

> The correct equation to begin with is:
>
> $$W = \int_{2}^{3} 3x\,dx + \int_{3}^{0} 4\,dy = 3\int_{2}^{3} x\,dx + 4\int_{3}^{0} dy$$
>
> $$W = 3\left[\frac{1}{2}x^2\right]_{2}^{3} + 4\left[y\right]_{3}^{0}$$
>
> $$W = \frac{3}{2}[3^2 - 2^2] + 4[0 - 3] = -45 \text{ J}$$

4. A plane is flying at a constant elevation of 1200 m with a constant speed of 430 km/h. At what angle ϕ should the cargo mate drop a package intended to land on

the ground at a specific point?

Answer:

$$\phi = \tan^{-1}\left(\frac{x}{y}\right) = \frac{1869}{1200}$$

The correct equation to begin with is:

$$y - y_0 = (v_0 \sin\theta_0)t - \tfrac{1}{2}gt^2$$

where $\theta_0 = 0$.

Solving for t will yield 15.65 s.
  Next the solution requires that the horizontal distance be calculated:

$$x - x_0 = (v_0 \cos\theta 0)t = 1869 \text{ m. If } x_0 = 0, \text{ then } x = 1869 \text{ m.}$$

The angle of sight is then:

$$V \approx 4.9 \text{ km/h}$$

---

5. Two skaters collide and embrace, which makes the collision a perfectly inelastic one. The mass and velocity of skater A is 83 kg, 6.2 km/h due east, and the mass and velocity of skater B is 55 kg, 7.8 km/h due north.

   a. What is the velocity of the two skaters together after the collision?
   b. What is the velocity of the center of mass of the two skaters before and after the collision?
   c. What is the fractional change in kinetic energy of the skaters because of the collision?

Answers:

a. $V \approx 4.9$ km/h

b. Same as for a.

c. 50 percent of the initial kinetic energy is lost as a result of the collision.

---

a. Linear momentum is conserved, so we can write:

$$m_A v_A = MV \cos\theta \quad m_A v_A = MV \cos\theta$$

This is the x component.

$$m_B v_B = MV \sin\theta$$

This is the y component.
  Now:

$$M = m_A + m_B$$

so that division yields:

$$\tan\theta = \frac{m_A v_A}{m_B v_B} = \frac{(55 \text{ kg})(7.8 \text{ km/h})}{(83 \text{ kg})(6.2 \text{ km/h})} = 0.834 \frac{\text{kgkm}}{\text{h}}$$

This gives

$$\theta \approx 40°$$

Also:

$$V = \frac{m_B v_B}{M \sin \theta}$$

which gives $V \approx 4.9$ km/h.

b. The velocity of the center of mass is the same as the velocity calculated for the two skaters together, because the center of mass (of the system) is not changed by the collision.

c. The initial kinetic energy of the skaters (the system) is:

$$KE_i = \frac{1}{2} m_A v_A^2 + \frac{1}{2} m_B v_B^2 = \left[\frac{1}{2}\right][83 \text{ kg}][6.2 \text{ km/h}]^2 + \left[\frac{1}{2}\right][55 \text{ kg}][6.2 \text{ km/h}]^2$$

$$KE_i = 3270\left(\frac{(\text{kg})(\text{km})^2}{(\text{h})^2}\right)$$

The final kinetic energy of the system is:

$$KE_f = \frac{1}{2} MV^2 = \left(\frac{1}{2}\right)(83 \text{ kg} + 55 \text{ kg})\left(4.86 \frac{\text{km}^2}{\text{h}}\right)$$

$$KE_f = 1630\left(\frac{(\text{kg})(\text{km})^2}{(\text{h})^2}\right)$$

The fractional change is then:

$$\text{frac} = \frac{K_f - K_i}{K_i}$$

$$\text{frac} = \frac{(1680 - 3270)\left(\frac{(\text{kg})(\text{km})^2}{(\text{h})^2}\right)}{3270\left(\frac{(\text{kg})(\text{km})^2}{(\text{h})^2}\right)}$$

$$\text{fraction} = -0.50.$$

This means that 50 percent of the initial kinetic energy is lost as a result of the collision.

CHAPTER 20

CATHOLIC UNIVERSITY OF AMERICA

PHYSICS 101 : TWENTIETH-CENTURY CONCEPTS OF THE PHYSICAL UNIVERSE

Charles J. Montrose, Associate Professor

AMIDTERM AND FINAL ARE GIVEN IN THIS COURSE. TOGETHER, THEY CONSTITUTE 60 percent of the course grade.

The object of the course is to introduce students to two of the major developments in twentieth-century physics, namely, Einstein's theory of relativity and the quantum theory. Like Caesar's Gaul, the course is divided in three parts. The first part covers some classical ideas about motion from the Greeks through Newton and some elements of Maxwell's electromagnetic theory. The focus is on establishing the principle of relativity (the states of rest and of uniform motion are indistinguishable). The second part of the course deals with the theory of relativity, the postulates on which it is based (and their experimental basis), and a few of their consequences. The final portion of the course introduces some of the ideas of the quantum theory. In this portion the approach is quasi-historical, using the nineteenth and early twentieth centuries' scientists' struggles with understanding the nature of atomic structure as the motivation.

Students should develop their analytical abilities so that they are able to apply general scientific principles to analyze specific situations. In many instances, the analysis must be quantitative, although no heavy-duty mathematics is required.

The examinations, somewhat to the chagrin of a not insignificant portion of the students, cannot be completed either by rote learning of class notes and chapter summaries (and then regurgitating them on the exams) or by memorizing a list of formulas and plugging in numbers to solve exam problems. The exam questions require students to *understand* what the basic laws of physics mean, to *apply* general laws, both singly and in combination, to analyze and predict the outcomes of observations under specific conditions.

There are no magic bullets. The road to success is an uphill one; otherwise, cliches such as "pinnacle of success" would make no sense. You have to work at it. But you're not stupid, and if you work sensibly, you will succeed.

Generally, don't let yourself get behind. Go to class. Once you're there, listen. Take *notes*—don't try to transcribe the lecture. Rewrite and annotate them. Keep up with the course on a class-by-class basis. Make sure that you understand Tuesday's class material before Thursday's class.

Be a pest; ask questions of the instructor and TAs (if any) both in class and during their office hours. Enlist the aid of a TA in an us-against-him (the instructor) strategy.

MIDTERM EXAM

Each question carries the same value—4 points. Because there are 25 questions, there are 100 possible points. Partial credit is given. For instance, in question 1, each of the parts, b through e, is worth 1 point.

Write the correct answers to the questions in the spaces provided. When it is appropriate, don't forget to give your answers units (e.g., pounds, seconds, Hz, and so on). In multiple-choice questions, more than one choice may be correct; give *all* the correct choices.

1. The formula developed by Newton for the gravitational force of attraction between two objects (masses = m and M) separated by a distance = r is

$$F = G\frac{mM}{r^2}$$

   For each of the following modifications to these conditions described in the left column, one of the consequences in the right column will result (generally *not* the one that is directly opposite it). Match them up in the spaces provided. As a sample, the first one, a, has been done for you.

a. If m is doubled, but everything else is left unchanged,   A. the force is not changed.

b. If m and M are both doubled, but r is not changed,   B. the force is doubled.

c. If m and M are not changed,
   but r is halved,

C. the force is halved.

d. If m and M are not changed,
   but r is doubled,

D. the force is quadrupled.

e. If m and r are doubled,
   but M is not changed,

E. the force is reduced
   by a factor of four.

Answers:

a. B

b. D

c. D

d. E

e. C

> The idea is to see if students understand what a formula is saying about the way a quantity (here the gravitational force) varies with each of the independent variables (here the masses and separation distance). Memorizing complicated formulas is not an objective of the course, and so the formula to be analyzed is given. For example, the students' understanding of what is meant by the *inverse square falloff of the force with separation* is tested in part c.

2. Make a sketch of the geocentric model of the universe introduced by Plato and his student Eudoxus and later refined by Ptolemy. Include on your sketch the sun, the moon, and the planets Earth, Mercury, Venus, and Mars. LABEL YOUR DRAWING CLEARLY.

   ◆ Demonstrate that you know what an epicycle is by including one on your sketch for the motion of Mars.

   ◆ *For a couple of extra points* explain what is meant by the *retrograde motion* of a planet.

Answer:

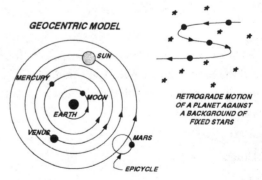

GEOCENTRIC MODEL

RETROGRADE MOTION
OF A PLANET AGAINST
A BACKGROUND OF
FIXED STARS

Circular orbits centered on the earth earn the student half-credit (two points) for the question. Arranging the sun, moon, and planets correctly gets one more point, and showing the epicycle correctly gets one more point. Explaining retrograde motion, either in words or by a sketch; gets two extra-credit points.

3. Sketch the heliocentric model proposed by Copernicus. Include on your sketch the sun, the moon, and the planets Mercury, Venus, Earth, and Mars.

   ◆ *For a couple of extra points* sketch in the trajectory of a comet such as, perhaps, the comet Hale-Bopp, which visited the solar system last year.

Answer:

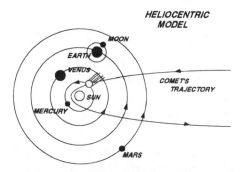

Whereas the ancients regarded the earth as unmoving (the geocentric universe of question 2), Copernicus (and, later, Kepler) demonstrated that a much simpler picture could be formulated if the sun was regarded as the focus with the planets orbiting about it. The point of covering this in the course is to begin to acquaint the students with examining phenomena from different reference frames in preparation for the theory of relativity. This question, along with question 2, tests the extent to which the students know what the geocentric and heliocentric pictures are. Orbits centered on the sun are worth two points, arranging the planets (Mercury, Venus, Earth, and Mars moving out from the sun) gets another, and the moon orbiting the earth gets the last point. Showing a highly eccentric orbit for a comet earns two extra-credit points.

4. In each of the sketches below, the dotted line gives the trajectory of an object whose position is shown at successive time instants labeled t_1, t_2, t_3, and t_4, separated by equal-time intervals of, say, 0.1 s. The arrows give the direction of the object's velocity at these time instants; at each instant, the arrow's length is proportional to the object's speed.

   For each of the cases illustrated, describe the force that operates on the object. Give as much detail as you can (e.g., direction and magnitude of force; variation of the force magnitude, if any; change in the force's direction, if any; and so forth).

   Hint: In answering this question, it may prove useful to redraw the sketches and to use additional arrows (or whatever) to indicate the force.

The solid arrows in the sketch were given in the question. The outlined arrows were drawn by the student in working out the response.

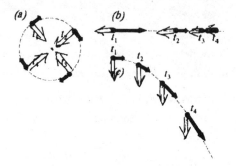

Answer:

The forces in each case are shown on the figure as unfilled arrows.

a. The force is always directed toward the center. The direction of the force changes, but its magnitude doesn't.

b. The force is directed toward the left: the object is slowed, stopped, and then accelerated in the opposite direction.

c. The force is directed downward: the object's horizontal velocity doesn't change, whereas its downward velocity increases.

The point of this question was to see if the students are able to apply what they should have learned about acceleration as the change in velocity and to connect this (via Newton's law) with forces. All three right earns four points, two right earns three points, and one right two points. None right is zero. Partial credit is given, and the total rounded to the nearest whole number one through four.

5. Which of the following are consequences of the Galilean relativity principle (which can be stated as: *The laws of mechanics are the same for all inertial observers*)?

a. If one inertial observer measures the velocity of a high-speed train to be 160 mi/h in a northeasterly direction, then all inertial observers will measure that train's velocity as 160 mi/h in a northeasterly direction.

b. All inertial observers will arrive at the same form for Newton's second law of motion, namely, force = mass × acceleration (or $F = ma$).

c. If one inertial observer measures the momentum of a complex system to have a magnitude of 1200 kg • m/s−1 and to be constant (not changing) in time, a second inertial observer need not obtain this same numerical value for the system's momentum, but must observe it to be nonchanging. (*Hint:* Momentum is defined as mass × velocity.)

   d. Motion with constant acceleration cannot be distinguished from the state of rest.

   e. The length of a stick and the time interval between two events will be the same for two inertial observers.

Answers: b, c

In effect, these are five true-false questions. Each is testing a different aspect of the students' understanding of the principle of relativity. (a) Students should realize that velocities differ for inertial observers in relative motion. (b) This is a *law of mechanics* and thus students should realize that it is the same for all inertial observers. (c) Conservation of momentum in the absence of forces is a *law of mechanics* and should be recognized as the same for all inertial observers. Because the velocities are different for observers in relative motion, the numerical value of momentum will differ. The phrasing tests if the students understand that for something to be *conserved* means that it is constant or not changing. (d) A state of constant acceleration can be distinguished from a state of rest in Galilean relativity. Does the student understand the ideas of inertia and uniform motion? (e) Although this is an assumption in classical mechanics, it does not follow from the Galilean relativity principle; indeed, it is not true in Einstein's theory of relativity, which is based on a principle that is more general and includes the Galilean one.

6. The circumference of a circle is given by $2\pi r$, where $\pi \approx 3.14$ and r is the radius. The mean distance from the earth to the sun is about 150,000,000 kilometers (km), and there are 31,500,000 seconds (s) in a year. The speed of the earth in its orbit around the sun is about

   a. 0.2 km/s

   b. 30 km/s

   c. 500 km/s

   d. 1000 km/s

   e. 30,000 km/s

The relevant equation is:

$$v = 2\pi r / T \sim 6.3 \times \frac{150{,}000{,}000 \text{ km}}{31{,}500{,}000 \text{ s}} \sim 30 \text{ km/s}$$

Answer: b

This is an elementary application of the definition of speed (= distance traveled divided by time elapsed). Students must recognize that in one year the earth travels a distance equal to the circumference of its orbit. A rough order-of-magnitude calculation is all that is needed, because the question is multiple-choice. The aim is to discover if the student can do an elementary calculation. Also, students are generally surprised to learn that as it moves in its orbit, the earth travels 30 km (about 19 mi) each second.

7. Astronauts in the space shuttle orbiting the earth experience weightlessness. However, if he were alive, Newton would say that they are falling. Explain why Newton would be correct.

Answer: They are accelerated toward the center of the earth as they move around it. As they "fall" toward the center of the earth, their motion around it keeps them a fixed distance away from its center and above its surface as shown in the sketch.

EARTH

Do students understand how it is that gravity is needed to maintain bodies in orbit? That without it, the object (the space shuttle) would fly off into space? The astronauts and their craft's free fall, owing to the gravitational attraction of the earth, is what keeps them and the craft in orbit. The key to a correct answer is appreciating the interplay between gravitational attraction and horizontal motion in producing stable orbits.

8. The definition of the average speed of an object is

$$\text{average speed} = \frac{\text{distance traveled}}{\text{time elapsed}}$$

In driving from Washington, D.C., to Hatboro, Pennsylvania, on the Wednesday before Thanksgiving, I was only able to average 65 kilometers per hour (km/h) for the journey, which took me 4 hours ($2\frac{1}{2}$ h on any other day). What is the Washington-to-Hatboro driving distance?

Answer:

$$\text{Distance} = (65 \text{ km/h}) \times (4 \text{ h}) = 260 \text{ km}$$

Question 8 (continued):

For an extra five points, what is my usual average speed?

Answer:

$$\text{speed} = \frac{260 \text{ km}}{2.5 \text{ h}} = 104 \text{ km/h}$$

Again, a simple application of the definition of the average speed of an object as distance traveled divided by time elapsed, in which the student must solve for the distance, knowing the time and the average speed. This is one step advanced from question 6, because this is not multiple-choice. The extra-credit part is a gift of five points for students who have done a similar homework question; thus, its purpose is to drive home the benefits of doing and reviewing homework assignments.

9. An observer at rest on the surface of the earth measures the acceleration of a sand-bag falling under the influence of gravity after being released by a hot-air balloon-ist who is rising at the steady speed of 10 m/s relative to the earth. Using Newton's second law ($F = ma$), the earth-based observer calculates the gravitational force on the object, obtaining the result $F = 27$ lb. If the balloonist were to determine the gravitational force on the sandbag, he would obtain a value for the force of gravity on the sandbag

 a. greater than 27 lb.
 b. less than 27 lb.
 c. equal to 27 lb.

Answer: c

Question 9 (continued):
For an extra few points, explain the basis for your answer to question 9.
Answer: Because both the earth-based observer and the balloonist are inertial observers, the principle of relativity implies that both must find that $F = ma$. Both will obtain the same values for m and thus for the weight.

> Does the student recognize that both the balloonist and earthbound observer are *inertial observers*? And that this implies $F = ma$ for both? And that they will get the same acceleration and mass for the sandbag? And thus the same gravitational force? Full credit for the correct answer to the multiple-choice and two extra points for correctly outlining the role of the principle of relativity in obtaining this answer.

10. One of the pitchers on CUA's baseball team is consistently able to throw a baseball so that it travels away from him at 90 mi/h. The following sketch depicts him deliv-ering such a pitch toward the east, while riding on a railroad flatcar that is travel-ing west at 40 mi/h. The catcher who will receive the ball is at rest on the earth. Under these conditions, the pitcher measures

 a. the ball's speed to be 90 mi/h,
 b. the ball's speed to be 130 mi/h,

 and the catcher measures

 c. the ball's speed to be 90 mi/h.
 d. the ball's speed to be 130 mi/h.
 e. the ball's speed to be 50 mi/h.

Answers: a, e

Simple applications of the classical velocity addition law for observers in relative motion. Students who miss this question usually need remedial work.

11. Electric forces
   a. are similar to gravitational forces in that their strength depends on the masses of the interacting objects.
   b. operate between bodies carrying electric charge.
   c. are inversely proportional to the square of the distance separating the charges.
   d. can either attract or repel electric charges.
   e. act between charges that are moving.

Answers: b, c, d, e

Five true-false questions about the nature of electric forces.

12. Magnetic forces
   a. can act on so-called nonmagnetic materials (e.g., copper wires).
   b. operate in such diverse applications as compasses and electric motors.
   c. can be produced by electric currents.
   d. can act on an electric charge that is moving.
   e. can act on an electric charge at rest.

Answers: a, b, c, d

Five true-false questions about the nature of magnetic forces.

13. Which of the following are true?
   a. Electric currents are the sources of both electric and magnetic fields.
   b. A current will be induced in a wire that is placed in a time-varying magnetic field.
   c. Except for gravitational forces, the only forces that we ever experience directly are electromagnetic in origin.
   d. A north magnetic pole is always accompanied by a south magnetic pole; similarly, a positive electric charge is always accompanied by a negative electric charge.
   e. Maxwell's theory of electromagnetism says that a time-varying electric field will cause a time-varying magnetic field and vice versa.

Answers: b, c, e

Five more true-false questions exploring some of the facts about electric and magnetic fields. Maxwell's equations in words.

14. Maxwell's theory of electricity and magnetism
    a. established that light propagated as waves and not as a stream of particles.
    b. contradicted Faraday's hypothesis regarding electromagnetic induction.
    c. predicted the existence of electromagnetic waves.
    d. suggested to him the existence of an all-pervading *aether* (or ether) as a medium in which light waves propagate.
    e. enabled him to calculate the speed at which electromagnetic waves propagate.

Answers: c, d, e

> What were some of the consequences of Maxwell's theory of electricity and magnetism? Five more true-false questions.

15. Newton's laws of dynamics tell us that if an observer wishes to calculate the motion of an object for all times after some initial time instant (call it t_0), he or she must know
    a. the object's mass.
    b. the object's location at the time instant t_0.
    c. the object's velocity at the time instant t_0.
    d. the total (net) force that acts on the body at the instant t_0 and at all times thereafter.
    e. his or her (i.e., the observer's) speed relative to the aether.

Answers: a, b, c, d

> What sort of things do Newton's laws of dynamics require for an input to calculate the motion of an object? Does Maxwell's luminiferous aether play any role?

The next three questions refer to the following sketch. It may be useful to recall that the speed of light in vacuum = 300,000 km/s.

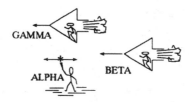

Alpha, Beta, and Gamma are three inertial observers. Just as he observes Gamma pass him, Alpha flashes a torch sending out two light flashes—one traveling to the left and one traveling to the right. At this same instant, Alpha observes that Beta is 200,000 km away and moving toward him. The sketch gives Alpha's view of the situation.

Questions 16 through 18 test the students' understanding of Einstein's principle that the speed of light is the same for all inertial observers and is $c = 300,000$ km/s. The next three questions also require that the students be able to examine a simple set of circumstances in different reference frames and come to some elementary conclusions about how the different inertial observers will perceive the motion—remembering that the velocity of material objects can never exceed c (a topic that is stressed in class as following directly from the principles of relativity and of the constancy of the speed of light).

16. When his own clock shows that 1 s has elapsed after he flashed the torch, Alpha notes that Beta is just passing him and Gamma is 150,000 km away from him. Alpha concludes that

    a. Gamma's speed is _____ km/s.
    b. Beta's speed is _____ km/s.
    c. the distances between Gamma and the two light flashes are _____ km and _____ km.

Answers:

a. 150,000 km/s.
b. 200,000 km/s.
c. 300,000 km and 300,000 km.

17. When Gamma's own stopwatch shows that 2 s have elapsed after Alpha passed him (Gamma observes Alpha moving to the right), Gamma will observe that the distance between himself and the

    a. light flash traveling to the left is 300,000 km.
    b. light flash traveling to the left is 600,000 km.
    c. light flash traveling to the right is 600,000 km.
    d. light flash traveling to the right is 900,000 km.

Answers: b, c

18. Which of the following are true?

    a. Gamma observes Alpha and Beta to be moving toward the right.
    b. Beta will measure the speeds of the light flashes as 150,000 km/s and 450,000 km/s.
    c. Beta will measure Alpha's velocity as 200,000 km/s toward the right.
    d. Beta will measure Gamma's velocity as 350,000 km/s toward the left.

Answer: c

19. As determined by an earth-based observer, a spaceship travels at 94.3 percent of the speed of light (at this speed $\gamma = 3$) from earth to a star system located 9 light-years from earth. The spaceship pilot will measure that

a. he ages by just over 3 years on the journey.

b. the distance between earth and the star system is 27 light-years.

c. the distance between earth and the star system is 3 light-years.

d. earth stay-at-homes age by a bit over 1 year during the journey.

e. earth stay-at-homes age by a bit over 9 years during the journey.

Answers: a, c, d

> This question explores whether the students understand the rules (ideas) of time dilation and length contraction in Einstein's special theory of relativity. The students must translate the given fact pattern from the earth-based observer's reference frame to that of the spaceship pilot and then apply the rules. The Lorentz (or Einstein) factor γ is given, so that memorizing the formula for it and doing the numerical calculations is not an issue.

20. Pions (represented by the symbol π) are radioactive particles that decay into a muon (symbol $= \mu$) and a neutrino (symbol $= \upsilon$). The decay is depicted schematically as $\pi \rightarrow \mu + \upsilon$. The half-life of pions measured by an observer at rest relative to them is 18 nanoseconds ($18 \text{ ns} = 18 \times 10^{-9} \text{ s} = 0.000000018 \text{ s}$). Consider a beam of pions traveling past two observers (Odie and Garfield) located 18 ft apart as in the following sketch. The speed of the pions is 99.5 percent of the speed of light (about 1 ft/ns) relative to the two observers. If Odie observes that 200 pions pass him each second, Garfield will observe that about

a. 200

b. 187

c. 100

d. 93

e. 20

pions pass him each second.

<table>
<tr><td colspan="3">Table of Einstein factors, γ</td></tr>
<tr><td>Velocity, v</td><td>v/c</td><td>$\gamma = \dfrac{1}{\sqrt{1 - (v/c)^2}}$</td></tr>
<tr><td>3,000 km/s</td><td>0.01</td><td>1.00005</td></tr>
<tr><td>30,000 km/s</td><td>0.1</td><td>1.0050</td></tr>
<tr><td>60,000 km/s</td><td>0.2</td><td>1.0206</td></tr>
<tr><td>120,000 km/s</td><td>0.4</td><td>1.0911</td></tr>
<tr><td>180,000 km/s</td><td>0.6</td><td>1.2500 ($= 5/4$)</td></tr>
<tr><td>240,000 km/s</td><td>0.8</td><td>1.6667 ($= 5/3$)</td></tr>
<tr><td>260,000 km/s</td><td>0.867</td><td>2.00</td></tr>
<tr><td>283,000 km/s</td><td>0.943</td><td>3.00</td></tr>
<tr><td>290,450 km/s</td><td>0.968</td><td>4.00</td></tr>
<tr><td>293,940 km/s</td><td>0.980</td><td>5.00</td></tr>
<tr><td>298,500 km/s</td><td>0.995</td><td>10.0</td></tr>
</table>

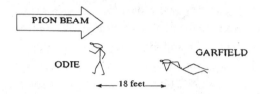

Answer: b

Two concepts come into play here—the idea of a radioactive half-life and the concept of time dilation. The situation is a recasting of the Smith-Frisch experiment discussed in class, which gave direct
experimental verification of the phenomenon of time dilation. Once the student recognizes that (as
measured by Odie and Garfield) a clock at rest with respect to the pions will run slowly compared
with their own clocks, the question should be easy. If there were no time dilation effect, half the
pions would decay in the Odie-to-Garfield journey and 100 pions per second (c) would be the correct choice. However, because the pion-clocks run slowly, fewer than half will decay meaning that
the number left will be somewhere between 100 and 200. The only choice that satisfies this is (b)
187. The idea here is to test whether the students understand elements that go into analyzing the
Smith-Frisch experiment, namely,

1. Using velocity and distance to obtain a time interval
2. The rule of time dilation
3. Radioactive decay half-life

well enough to analyze an analogous situation.

21. An enchanted prince has been turned into a frog and can turn back into a prince
    only if he is kissed by a beautiful princess on his 21st birthday. The only beautiful
    princess around is exactly 2 years younger than the prince, and is not allowed to
    look upon a man until she is 21. By making a high-speed round-trip journey on a
    high-speed spaceship, the frog-prince can arrange to age less rapidly than the
    princess.

    a. If he leaves on his 20th birthday (which happens to be her 18th birthday), at
       what speed must he travel so that he returns on his 21st birthday, just as the
       princess is also celebrating her 21st birthday. (*Hint:* First decide what γ must be;
       then work out the speed.)

Answer: $0.943c$ (or 283,000 km/s)

    b. How far will he have traveled (as measured by the princess)?
       A. less than 1.0 light-year
       B. 1.0 light-year
       C. 2.8 light-years
       D. 3.0 light-years
       E. 3.2 light-years

Answer: C

I think I stole this question from Sheldon Glashow's book, *From Alchemy to Quarks*. This question tests whether students can use the table of Einstein factors to analyze a given fact pattern in which time dilation is the all-important idea. The student must also realize that by moving at a speed just less than the speed of light for 3 years, the traveler will cover a distance just under 3 light-years. Hence, he or she must understand what a light-year is.

22. A pole-vaulter attempting to set a new (if unofficial) world's record decides to use an unusually long approach to build up speed for her vault. On the approach path that she follows is located the storage shed in which the track team stores its equipment. The vaulter decides to open the front and rear doors of the shed so that she can run through the shed undisturbed. She also notes that when she lays it down, her pole just spans the distance between the doors. On her approach run, the vaulter, who is carrying her pole horizontally, reaches a speed of 260,000 km/s (!! ??).

a. Place the following events in their proper sequence as observed by the pole-vaulter.

A. The front of the pole is adjacent to the shed entrance.
B. The rear of the pole is adjacent to the shed entrance.
C. The front of the pole is adjacent to the shed exit.
D. The rear of the pole is adjacent to the shed exit.

Answer:
First: A
Second: C
Third: B
Fourth: D

b. For this situation, sketch the world lines (the paths through space-time) on the following Minkowski space-time diagram for

1. the shed entrance
2. the shed exit
3. the front of the pole
4. the rear of the pole

from the reference frame of the track coach who is at rest relative to the shed. Label each path clearly as 1, 2, 3, and 4.

Answer:

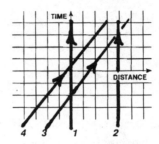

The question tests the students' understanding of length contraction in the theory of relativity. Can the student transform from the "natural" earth-based frame of reference to the reference frame of the pole-vaulter?

23. Einstein's famous mass-energy formula:

$$E = mc^2$$

    a. enables us to calculate the energy of a body of mass m moving at a speed = c.

    b. means that we can compute the amount of work that a body can do if we know its mass.

    c. asserts that energy can neither be created nor destroyed.

    d. means that when a firefly emits light, its mass decreases.

    e. asserts that inertia (i.e., mass) and energy are the same thing.

Answers: b, c, d

Five true-false questions on Einstein's mass-energy formula $E = mc^2$ that test the students' understanding of what this equation means.

24. Two *globs* of putty (each of rest mass 10 g) collide head-on, stick together (forming a *blob*), and stop. Before colliding, each glob was traveling at about 283,000 km/s, so that each had a precollision relativistic mass of 30 g. Just after the collision, the rest mass of the *blob* is:

    a. 20 g

    b. 30 g

    c. 40 g

    d. 60 g

    e. 80 g

Answer: d

Another exploration of the mass-energy equivalence. This question is quite sophisticated in that, in addition to understanding $E = mc^2$, the student must be able to apply the laws of conservation of momentum and energy. The "obvious" (and incorrect answer) of 20 g (a) was chosen by just under half the students. The correct answer, 60 g (d), was selected by slightly fewer students.

25. When an electron with energy 0.8 MeV (0.8 million electronvolts) is brought to rest, perhaps by colliding with a lead brick, its energy becomes its rest mass energy, which is about 0.5 MeV. The 0.3 MeV of energy that is given up by the electron (the rest of the energy that it had)

a. goes into some other form of energy, perhaps heat or light.

b. goes into increasing the electron's mass.

c. goes into raising the electron's temperature.

d. is simply lost.

Answer: a

Energy conservation and the conversion of energy from one form to another are the ideas that students must be able to apply to answer this question correctly.

FINAL EXAM

Each question carries the same value—4 points. Because there are 25 of them, there are 100 possible points. Partial credit is given. For instance, in question 1, each of the parts is worth 1 point. The first 13 questions test the material covered in the first 60 percent of the course (classical physics and relativity); most of this was tested on the midterm exam, and a few of the questions are near-repeats from that exam. The last 12 questions cover the atomic physics and quantum physics part of the course—roughly the last 40 percent.

Specifically, in preparing for exams, do the assigned homework questions yourself after discussing them with a small (two- to three-person) study group. Use the homework questions as a guide to prepare for the exams. They usually tell you what the instructor thinks are the main topics.

If the instructor makes available his lecture notes, emphasize them in your studying. Pay attention to examples that the instructor works out in class. If the instructor says about a specific example, "Something like this will be on the exam," believe it.

Use old exams as a guide. (I am always astonished at how many students ignore this advice, which I always give.)

In taking the exams, READ THE QUESTION. Then, read it again. Then think about what it means. If you can, draw a sketch or sketches of the situation to which the situation refers. Don't skip these steps in a rush to answer the question. You get very little credit just for answering a question. Credit is given for answering it correctly. After you have decided on your answer, reread the question to make sure that you are answering what is asked.

Please write the correct answers in the spaces provided. In the multiple-choice questions, more than one answer may be correct; be sure to give *all* the correct choices. (At least one answer will be correct.)

1. Newton's formula for the gravitational force of attraction between two objects (masses = *m* and *M*) separated by a distance = *r* is:

$$F = G\frac{mM}{r^2}$$

a. According to this, the gravitational force is not changed if *m*, *M*, and *r* are all doubled.

b. This formula is consistent with Kepler's law that the planets move in elliptical orbits with the sun at one of the foci.

c. The gravitational attraction between the earth and the moon explains why the moon is held in orbit around the earth, rather than wandering off into space.

d. This is similar to the electrical force of attraction between a positive and a negative charge in that both forces are inversely proportional to the square of the separation of the interacting objects.

Answers: b, c, d

Basically four true-false questions. In choice a, the idea is to see if students understand what a formula is saying about the way a quantity (here, the gravitational force) varies with each of the independent variables (here, the masses and separation distance). Choices b and c test whether students appreciate the role of the gravitational force in orbital motion. Choice d appears as a reminder of the origins of the planetary model of the atom.

2. Each of the following sketches shows the trajectory of an object. The positions of the object are shown at a starting instant (labeled "0"), and then at time instants 1, 2, and 3 s after this. The shaded arrows give the direction of the object's velocity at each instant and the length of the arrows is proportional to the object's speed.

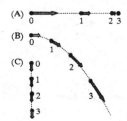

a. In (A), the object is acted upon by a constant force directed toward the left.

b. In (B), the object is acted upon by a constant force directed downward.

c. In (B), the object is acted upon by a force that is constant in magnitude, but whose direction is changing.

    d. In (B), the object is acted upon by a force with a constant horizontal compo-
        nent and with an increasing downward component.

    e. In (C), the object is acted upon by a uniform force directed downward.

Answers: a, b

> This is nearly a repeat from the midterm exam. The point is to ensure that students have learned about acceleration as the change in velocity and to connect this (via Newton's law) with forces. All three right earns four points, two right earns three points, and one right two points. None right is zero. Partial credit is given.

3. The moon is approximately 375,000 km from the earth and a bit under 700 h. Its orbital speed around the earth is about

    a. 0.35 km/h.

    b. 35 km/h.

    c. 3500 km/h.

    d. 350,000 km/h.

> The relevant equation is:
>
> $$v = 2\pi r / T \sim 6.3 \times \frac{375{,}000 \text{ km}}{700 \text{ hr}} \sim 3500 \text{ km/hr}$$

Answer: c

> A variation on a question that appeared on the midterm. This is a straightforward application of the definition of speed (= distance traveled divided by time elapsed). Students must recognize that in one full orbit around the earth, the moon travels a distance equal to 2π times the radius of its orbit, which is the earth-moon distance. A rough order-of-magnitude calculation is all that is needed, because the question is multiple-choice.

4. A passenger at rest on an eastbound train traveling at a speed of 60 mi/h drops his favorite steel marble. Which of the "multi-flash" pictures sketched below (taken at equal time intervals of 1/20 s) best represents the trajectory of the marble as seen by

    a. the passenger?

    b. a hobo on the ground watching the train pass?

    c. the driver of an eastbound car traveling at a speed of 90 mi/h?

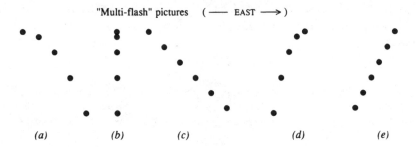

"Multi-flash" pictures   (—— EAST ——→)

(a)   (b)   (c)   (d)   (e)

Answers:

a. b

b. a

c. d

This is an opportunity for students to persuade me that they do recognize the differences and similarities in observations of a simple sequence of events made by different inertial observers.

5. Suppose Mr. Lucky is driving about 100 ft behind a truck loaded with cannonballs. Both are traveling at a steady speed of 55 mi/h. At this speed Mr. Lucky's car requires about 250 ft to come to a stop after he applies the brakes. Just after both vehicles have driven onto a 1000-ft-long single-lane bridge that is 80 ft above a churning, boulder-strewn river, several dozen cannonballs drop off the rear of the truck. When they hit the road, they begin bouncing up to heights of between 5 and 10 ft. Mr. Lucky's best option under these circumstances is to

   a. swerve off the bridge and take his chances in the river, because he cannot stop in time.

   b. jump out of his car and dive into the river.

   c. hit the brakes and hope for the best.

   d. pray; only a miracle can help him at this point.

   e. speed up—he may as well get it over with.

Answer: c

The idea is to test the students' appreciation of Newton's first law. Just because the cannonballs fall off the truck does not mean that they have lost their 55-mi/h forward velocity. Apart from their gradual slowing from air resistance and friction with the road when they bounce, the balls will continue to move with this velocity. Mr. Lucky will have plenty of time to apply the brakes and stop in safety. (This question describes a situation that happened to me some 25 years ago while driving in Massachusetts—the objects were concrete blocks rather than cannonballs, and it was on an interstate highway rather than a one-lane bridge, but the traffic was such that there was no opportunity to change lanes.)

6. A major-league pitcher is consistently able to throw a baseball so that it travels away from him at 90 mi/h. The following sketch depicts him delivering such a pitch while riding on a railroad flatcar that is traveling at 60 mi/h as shown. The catcher who will receive the ball is at rest on the earth. Under these conditions, the pitcher measures

a. the ball's speed to be 30 mi/h,
b. the ball's speed to be 90 mi/h,

and the catcher measures

c. the ball's speed to be 30 mi/h.
d. the ball's speed to be 90 mi/h.
e. the ball's speed to be 150 mi/h.

Answers: b, c

A repeat from the midterm, this should be a gift question. Over 80 percent of the students got it right then. Just straightforward applications of the classical "velocity addition law" for observers in relative motion.

7. Which of the following are true?
  a. Magnetic fields act on electric currents; electric fields act on electric charges.
  b. Maxwell's theory of electromagnetism says that a time-varying magnetic field will cause a time-varying electric field and vice versa.
  c. Because of electromagnetic forces, two objects cannot occupy the same space at one time.
  d. Maxwell's theory of electromagnetism enabled him to calculate the speed at which electromagnetic waves travel without mentioning any specific reference frame. Maxwell added the idea of the *aether* to provide a propagation medium for the waves.
  e. Infrared, microwaves, ultrasonic waves, and x-rays are all types of electromagnetic waves.

Answers: a, b, c, d

Five true-false questions about the electromagnetic theory and electromagnetic waves. Choice b is a bit ambiguous because it is possible for a time-varying magnetic field, say, to induce a *constant* electric field (if the magnetic field strength increases linearly with time). I have asked this question of over 400 students; none has ever been confused by this subtlety. This is because I cover this topic in connection with the topic of electromagnetic waves. If a student ever brought the ambiguity to my attention, I would probably give him or her a substantial bonus on the exam.

Questions 8 through 13 refer to the following sketched situation (NOT TO SCALE).

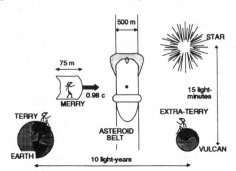

The sketch displays the reference frame of Terry, at rest on the earth, as he observes a spaceship piloted by Merry at a speed of 98 percent of c on a one-way trip from earth to the planet Vulcan. Vulcan, the star, and the asteroid belt are all at rest relative to Terry and to Extra-Terry on Vulcan. At $\upsilon = 0.98c$, the Einstein factor is $\gamma = 5$. The measurements shown in the sketch are those made by Terry:

Earth-to-Vulcan distance = 10 light-years
Vulcan-to-star distance = 15 light-minutes
Length of Merry's spaceship = 75 m
Width of asteroid belt = 500 m

Questions 8 through 13 concern relativistic kinematics. Question 8 explores whether the students can apply the rule (idea) of length contraction in Einstein's special theory of relativity. The students must translate the given fact pattern from the earth-based observer's reference frame to that of the spaceship pilot and then apply the rule. The Lorentz (or Einstein) factor γ is given so that memorizing the formula for it and doing the numerical calculations is not an issue.

Question 9 tests whether the students can apply the rule (idea) of time dilation. In this question it is also necessary to realize that the earth-based observer will determine that the duration of a 10-light-year journey at a speed of 98 percent of c is a bit over 10 years.

Question 10 tests the notion of clock synchronization in the theory of relativity. What is the meaning of simultaneity?

Question 11 tests whether the students can apply the rule of length contraction. This is a repeat of an idea from the midterm, in which students had trouble and which was subsequently reexamined in class.

Question 12, on the face of it, is just another length contraction question. However, students sometimes use the principle that all inertial observers measure the speed of light to be c, with the notion that the time interval for light to travel between two locations must be the same. I was pleased that over 75 percent of the students got this question right (at the same time, I was not happy that almost 25 percent missed it).

On the midterm, there was a similar question in which students were asked to draw the world lines for several entities moving through space-time. Fewer than 25 percent of the students correctly answered that question. We went through the ideas in class after that exam, and the class was told that there would be a similar question on the final. Question 13 was it, and this time nearly 60 percent got it right. Still 40 percent either didn't believe me, didn't care, or didn't figure it out and didn't ask.

8. a. What will Merry measure for the width of the asteroid belt?

b. What will Merry measure for the length of her spaceship?

c. What will Merry measure for the distance from Vulcan to the star?

Answers:

a. 100 m

b. 375 m

c. 15 light-minutes

9. According to Merry, she will age by (question 9a) during her earth-to-Vulcan journey, while her brother Terry will age by (question 9b) during the same time.

a. 1.96 years

b. 2.04 years

c. 5.00 years

d. 9.80 years

e. 10.20 years

Answer to 9a: b

Answer to 9b: e

10. Terry and Extra-Terry have synchronized their clocks. Merry will

a. also observe that they are synchronized.

b. observe Terry's clock to be 9.8 years behind Extra-Terry's.

c. observe Terry's clock to be 9.8 years ahead of Extra-Terry's.

d. observe Terry's clock to be 2 years behind Extra-Terry's.

e. observe Terry's clock to be 2 years ahead of Extra-Terry's.

Answer: b

11. Place the following two events in their proper sequence:

(a) Front of the spaceship emerges from the asteroid belt

(b) Rear of the spaceship enters the asteroid belt

    (i) ... according to Terry: first _____   second _____

    (ii) ... according to Merry: first _____   second _____

Answers:

(i) ... according to Terry: first: b   second: a

(ii) ... according to Merry: first: a   second: b

12. What does Merry measure for the time it takes a light flash to travel from Earth to Vulcan?

Answer: 2 years

13. On the Minkowski space-time diagram that follows, sketch the world lines (the paths through space-time) according to Merry of

   a. the spaceship
   b. Earth
   c. Extra-Terry
   d. a light flash sent from Earth to Vulcan

   Use the following scales:

   time axis: 1 year = 2 divisions
   distance axis: 1 light-year = 2 divisions

The following sketch incorporates the student's answer.

Answer:

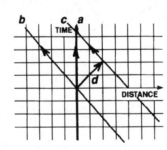

14. A carbon atom is 12 times as heavy as a hydrogen atom. Methane is a compound composed of carbon and hydrogen in which the total amount of carbon weighs three times as much as the total amount of hydrogen. The chemical formula of methane is

   a. CH_3
   b. C_3H
   c. CH_4
   d. C_6H_2

Answer: c

This question examines the students' understanding of how chemical formulas are put together if the relative weights of the constituents are known. This idea was discussed in connection with the atomistic concepts developed by Dalton and the nineteenth-century chemists.

15. a. Make a sketch of the so-called "plum-pudding," or "raisin-cake," model of an atom. Label the main features.

Answer:

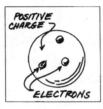

15. b. Weaknesses of this model include:

    a. its inability to account for atoms' electrical neutrality.

    b. its lack of stability—the electrons spiraled to the center in about 108 s.

    c. its inability to explain atomic line spectra.

    d. its failure to explain Rutherford's alpha-particle scattering results.

Answers: c, d

What was Thompson's "plum-pudding" atomic model, and what were its shortcomings that led to its being discarded?

16. The following figure shows the allowed energy levels of a hydrogen atom [energies in electronvolts (eV)]. Transitions to the $n = 1$ state are called *Lyman transitions*.

    What are the energies of the two lowest-energy photons that can be emitted in *Lyman transitions*?

$$
\begin{array}{ll}
0\ \text{eV} & n = \infty \\
-0.85\ \text{eV} & n = 4 \\
-1.5\ \text{eV} & n = 3 \\
\\
-3.4\ \text{eV} & n = 2 \\
\\
\\
\\
-13.56\ \text{eV} & n = 1
\end{array}
$$

ENERGY ↑

Answers:

12.06 eV and 10.16 eV

This question tests whether students understand the connection between the specific discrete energy level structure of an atom and the energies of photons that can be emitted by the atom. Although a similar question was asked on an earlier test and on a homework assignment, still nearly half the class missed it. The most common wrong answers were –3.4 eV and –13.56 eV.

17. Below are two electromagnetic waves. In the sketches, the distance between the vertical "tick" marks is 500 nm.

   a. Estimate the wavelength of the wave at the top.
   b. Estimate the wavelength of the wave at the bottom.
   c. The frequency of the wave at the top is 450 THz; a reasonable estimate of the frequency of the wave at the bottom is:

   a. 650 THz
   b. 450 THz
   c. 250 THz

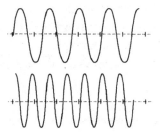

   Answers:
   a. 650 nm
   b. 450 nm
   c. a

Do the students understand what wavelength is? Can they read it from a sketch for a sinusoidal wave? In question 17c, the students' grasp of the inverse relationship between wavelength and frequency is tested.

18. The following sketch shows the "black-body radiation" spectrum emitted by an object at a temperature of 3000°.

   a. On the diagram, sketch the spectrum emitted by an object at 6000°.
   b. The 6000° object is white-hot. What will be the color of the 3000° object?

Answers:

a.

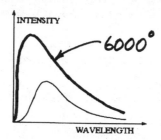

b. Red

How does the black-body spectrum depend on temperature? Three elements had to be present in the sketch to get full credit (3 points) for part a: the 6000° spectrum had to be drawn with
  a. a larger maximum than the 3000° spectrum
  b. the maximum located at a shorter wavelength
  c. the entire 6000° spectrum above the 3000° spectrum

Part b was worth one point; orange, red-orange, and red were all acceptable answers.

19. Planck "explained" the black-body radiation spectrum by hypothesizing that the atomic oscillators that emit the radiation could exist only in equally spaced discrete energy levels:

Energy = nhf

where $n = 0, 1, 2, \ldots$, h = Planck's constant, and f = the oscillator frequency.

Following, the first diagram shows the energy levels for an oscillator with a frequency of 480 THz. On the other two diagrams, sketch the energy levels for oscillators of frequencies 360 THz and 960 THz.

Answer:

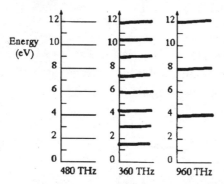

Do the students understand the concept that the energy level spacing is proportional to the oscillator frequency? Even some who got it right did so by having memorized Planck's constant and carrying out the multiplications numerically on their calculators. Most students did something intelligent with this question, but about a quarter were completely mystified.

20. Rutherford's experiments on the scattering of alpha particles by a thin gold foil established

   a. that most of the alpha particles passed through the gold foil either not deflected or only slightly deflected from a straight-line path.

   b. that there were too many alpha particles backscattered (bounced back) by the gold foil to be explained by the so-called plum pudding model of atomic structure.

   c. that the half-life of nuclei undergoing alpha-particle decay can exceed 1 billion years.

   d. that gamma rays are electromagnetic waves with wavelengths shorter than X rays.

   e. the existence of the atomic nucleus.

Answers: a, b, e

What were the experimental observations and what implications were drawn from Rutherford's alpha particle–scattering experiment? We spent a lot of time describing and discussing this in class, and it is treated in some detail in the book as well as in notes that I distributed to the students.

21. a. De Broglie's formula connecting a particle property (momentum) and a wave property (wavelength) is:

   a. $p = h\lambda$
   b. $p = h/\lambda$
   c. $n\lambda = 2\pi p \ (n = 1, 2, 3, \ldots)$
   d. $p = \lambda/h$

   b. De Broglie used his hypothesis that electrons have a wavelike nature to explain why only certain orbits are possible in a hydrogen atom. Sketch the de Broglie wave for the $n = 3$ state of hydrogen.

Answers:

a. a

b.

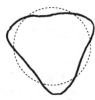

De Broglie's famous formula connecting momentum and wavelength was a major theme in our discussion of the quantum theory. His use of it to explain Bohr's orbit condition

angular momentum = integer × Planck's contant/2π

was treated, and sketches similar to that required were drawn in class.

Question 21b tested whether the students understood that only when the wave exactly closes on itself is the orbit allowed.

22. The following chart shows the dependence of the kinetic energy of electrons ejected from an electron metallic surface on the frequency of the light kinetic (electromagnetic waves) illuminating the metal. For the case shown, the electron binding energy is 4 eV.

On the chart, sketch the graph of electron kinetic energy versus incident light frequency for a metal for which the binding energy is 6 eV.

Answer:

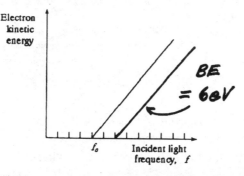

Three elements were necessary to receive full credit for this question. The required graph had to be

1. A straight line
2. With the same slope as the given line
3. With a frequency-axis intercept shifted toward higher frequencies from the given line

Under test on this question was mainly the graphical representation of the formula given in the hint for question 25 and the connection between critical frequency and binding energy.

23. Following (see Answer b) is a sketch of ψ^2 (ψ is the wave function) for an electron traveling to the right (the steeper wave).

a. What is the uncertainty in position, Δx, for the electron at the instant shown?
b. Suppose that during the next femtosecond (a femtosecond is one-millionth of a nanosecond), the electron moves 1 nm to the right. On the graph, make a sketch of ψ^2 at this later instant.

Answers:

a. 0.4 nm

b.

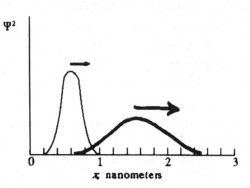

(The answer is the wave to the right.)

> **Does the student understand the ideas of what Δx means and a spreading wave packet representing the motion of a free particle?**
> In part a, I accepted anything between 0.2 and 0.8 nm as being correct, although 0.4 nm (the full width of the packet at about half-height) was what corresponded to our classroom definition.
> In part b, I wanted to see a broadened pulse of lower peak height with the peak position shifted to the right to a location about 1.6 nm.

24. Heisenberg's uncertainty principle, $\Delta x \, \Delta p \geq \frac{1}{2} h$,

   a. describes the intrinsic lack of determinism in nature.

   b. renders impossible the Bohr idea of electrons moving in circular orbits about a nucleus.

   c. requires that physical entities have both a wave and a particle aspect to their nature.

   d. asserts that any attempt to improve the accuracy of a measurement of an object's position will result in an increase in the particle's momentum.

Answers: a, b, c

> **What does the uncertainty principle say and what does it not say? These are just four true-false questions.**

25. In the photoelectric effect using the metal cesium, red light (wavelength = 660 nm) will just free an electron from the metal. The entire energy of the photon is used up in overcoming the attractive forces that bind the electron in the metal; there is no energy left over after the electron has escaped. If green light (wavelength = 550 nm) is used, what fraction or percent of the photon's energy does the electron retain as its kinetic energy after it has escaped from the cesium? *Hints:* If you do a

bit of algebra before plugging in the numbers, you will avoid a lot of complicated computations. Some of the following formulas may be useful:

$$\epsilon = \hbar f$$

$$c = f\lambda$$

$$KE(electron) = \hbar f - BE$$

Answer:

Let $\lambda_0 = 660nm$; then $f_0 = c / \lambda_0$ and $BE = hc / \lambda_0$.

For $\lambda = 550nm, KE = hc / \lambda - hc / \lambda_0$ and $\dfrac{KE}{hc / \lambda} = 1 - \dfrac{hc / \lambda_0}{hc / \lambda} = 1 - \dfrac{\lambda}{\lambda_0} = 1 - \dfrac{5}{6} = \dfrac{1}{6}$

$\frac{1}{6} = 16\frac{2}{3}\%$

One problem involving a detailed calculation. I promised (threatened?) the students that there would be one such question on the final. Only about 20 percent got it completely right, but another 30 percent got at least half credit. But, there were almost 40 percent who did nothing correct, often nothing at all. Probably too hard a question, but some students had told me that the exams would be easier if they involved more numerical problems (they implied that many of their fellow students felt that way). By and large, the students who got this question right were the ones who got As and Bs in the course anyway.

CHAPTER 21

PHYSICS 107: INTRODUCTORY PHYSICS

Christopher D. Wentworth, Associate Professor

W E GIVE 12 UNIT EXAMS (30 MINUTES EACH) AND A COMPREHENSIVE FINAL. AN example of 1 unit exam and a final are included here.

Students who complete the course will

1. Be able to describe and formulate problems about the physical world in a way that allows quantitative analysis in the chosen topic areas

2. Have gained experience in making and describing observations in the area of mechanics

3. Have gained experience in developing hypotheses concerning mechanics observations

4. Have gained experience in organizing experimental data so that it can be analyzed in a precise and quantitative manner

5. Be able to use the computer as a data acquisition and analysis device

6. Have gained experience in working with a group to achieve intellectually complex goals

We want students to recognize how basic physics concepts are involved in everyday phenomena. So, given a novel situation involving everyday phenomena, we would like our students to recognize and describe as many physics principles as possible that are evident in the situation. We would like students to explain why they think a concept applies and why people believe in the particular concept. Important questions are "How do we know that is true?" or "Why do we believe that?" Saying that a textbook says it is true, or that the instructor says that it is true is generally not adequate.

Given a physical situation using everyday objects, we would like our students to make predictions about what will happen to the system, or at least recognize what information is needed to enable predictions to be made.

We want our students to interpret data and analyze it quantitatively.

Most of the problems included in our exams deal with situations that the student has seen in laboratory exercises or that should be a part of everyday experience. There is usually some new twist to the situation, so that the student is not just reporting back memorized facts. We almost always request that a student give some explanation for a result.

Many of our problems give data in either table or graphical form. Students must interpret the data or analyze it appropriately to answer the questions.

MIDTERM (UNIT) EXAM

1. Consider the following argument between two students. With which student do you most agree? Explain why you made your choice.

   Student A: We have made these 10 time measurements, and I have calculated an average of 42.1 s and a standard deviation of 0.3 s. Therefore, I would estimate the uncertainty in my average due to random variations is about 0.3 s.

   Student B: Well, I disagree with you. The standard deviation may tell you about the uncertainty in the data points, but not about the uncertainty in the average. We need to do something else to find the uncertainty in the average due to random variations.

Answer: Student B is correct.

The standard deviation tells us about the uncertainty in individual measurements.

The standard deviation of the mean tells us about the uncertainty in the calculated average.

If we repeat the series of 10 measurements, we will likely get a different calculated average, but the variation tends to be less than the variation of the individual data points.

> Strategy for solution: First, I identify the subject area of the question. It is concerned with error or uncertainty analysis of experimental data. Now, I try to recall some major points that we discussed about this subject. Generally, two things can happen when we take measurements: there can be a systematic error introduced into the measurements, and there can be random variations in each measurement. We associate the word "uncertainty" with the random variations. The mathematical theory of statistics lets us make some estimates in how much uncertainty there is in a sample of

repeated measurements due to random variation. The standard deviation tells us something about the uncertainty in individual measurements, and the standard deviation of the mean tells us about the uncertainty in the calculated average. These are the relevant facts to consider in analyzing this situation.

2. A person walks in front of an ultrasonic motion detector. The position-time graph is shown. Draw the corresponding velocity-time graph. Identify the positions A, B, and C on the velocity graph.

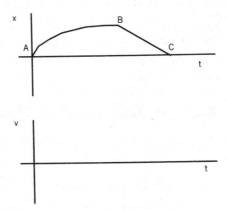

Describe in words the motion of the person in the indicated regions:

A to B:
 Direction of travel?
 What is happening to the speed?
 Is acceleration positive, negative, or zero?

B to C:
 Direction of travel?
 What is happening to the speed?
 Is acceleration positive, negative, or zero?

Strategy for solution: First, I identify that this problem is concerned with describing motion—kinematics. Now I recall some important points about describing motion with graphs. The velocity-time graph tells us something about the shape of the position-time graph. Generally, the velocity is the slope of the line drawn tangent to the position-time graph at a particular point. So, if the position-time graph is straight, the velocity will be constant (horizontal) for that time. If the slope of the tangent line to the position-time graph is changing, then the velocity must be changing. Finally, the sign of the velocity tells us something about the direction of travel. The acceleration-time graph tells us something about the shape of the velocity-time graph.

Answer:

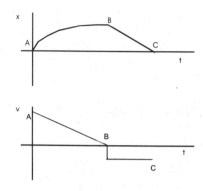

A to B:

   Direction of travel? Positive x direction, away from the origin.

   What is happening to the speed? It is decreasing.

   Is acceleration positive, negative, or zero? Negative.

B to C:

   Direction of travel? In the negative x direction, toward the origin.

   What is happening to the speed? It is constant.

   Is acceleration positive, negative, or zero? Zero.

3. The following table gives some position-time data for a ball rolling on a gym floor. Make estimates of the instantaneous velocity and the instantaneous acceleration at 1-s intervals.

t (s)	x (m)	v_x (m/s)	a_x (m/s²)	y (m)	v_y (m/s)	a_y (m/s²)
0.0	0.0			0.0		
1.0	1.2			0.3		
2.0	2.0			1.2		
3.0	2.4			2.7		
4.0	2.4			4.8		
5.0	2.0			7.5		
6.0	1.2			10.8		

Strategy for solution: First, I identify that there are two topic areas involved here: (1) the definition of instantaneous velocity and acceleration and their relationship to position-time data and (2) describing two-dimensional motion. Now, I recall some important facts about these topics: two-dimensional motion can be expressed in terms of separate/independent descriptions of motion in the x direction and in the y direction. Also, instantaneous velocity can be approximated from average velocity calculated over a short time period, and instantaneous velocity can be approximated from average acceleration over a short time period.

   So, we can estimate the instantaneous x component of velocity at $t = 0.5$ s by calculating the average velocity from $t = 0.0$ to 1.0 s.

$$v_x(0.5\ \text{s}) \cong \frac{1.2\ \text{m} - 0.0\ \text{m}}{1.0\ \text{s} - 0.0\ \text{s}} = 1.2\ \text{m/s}$$

and the instantaneous velocity at $t = 1.5$ s is estimated by calculating average velocity from $t = 1.0$ to 2.0 s.

$$v_x(1.5\ \text{s}) = \frac{2.0\ \text{m} - 1.2\ \text{m}}{2.0\ \text{s} - 1.0\ \text{s}} = 0.8\ \text{m/s}$$

and so on.

Similarly, the instantaneous x component of acceleration at $t = 1.0$ s is estimated by calculating the average acceleration from $t = 0.5$ to $t = 1.5$ s.

$$a_x(1.0\ \text{s}) = \frac{0.8\ \text{m/s} - 1.2\ \text{m/s}}{1.5\ \text{s} - 0.5\ \text{s}} = -0.4\ \text{m/s}^2$$

Answer:

t (s)	x (m)	v_x (m/s)	a_x (m/s^2)	y (m)	v_y (m/s)	a_y (m/s^2)
0.0	0.0			0.0		
0.5		1.2			0.3	
1.0	1.2		-0.4	0.3		0.6
1.5		0.8			0.9	
2.0	2.0		-0.4	1.2		0.6
2.5		0.4			1.5	
3.0	2.4		-0.4	2.7		0.6
3.5		0.0			2.1	
4.0	2.4		-0.4	4.8		0.6
4.5		-0.4			2.7	
5.0	2.0		-0.4	7.5		0.6
5.5		-0.8			3.3	
6.0	1.2			10.8		

4. Estimate the x and y components of the velocity vector at $t = 0$ s. Explain how you made the estimate.

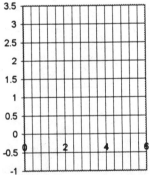

Answer:

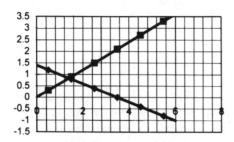

The line for v_x hits $t = 0$ s at about $v_x = 1.4$ m/s.
The line for v_y hits $t = 0$ s at about $v_y = 0.0$ m/s.

5. If the acceleration in the preceding situation is being caused by two people tapping on the ball, then describe how they must be doing the tapping to give this motion.

> **Strategy for solution:** I must remember that in describing two-dimensional motion, the x and y directions can be considered independently. So, one picture here is that one person taps in the x direction to cause acceleration in that direction, and the other person taps in the y direction to cause acceleration in that direction. Acceleration is a vector, so direction of tapping is important. We determine this by looking at the slope of the velocity lines.

Answer:

Because the slope of the v_x line is negative, the acceleration in the x direction must be negative. So, one person must be tapping in the minus x direction.

Because the slope of the v_y line is positive, the acceleration in the y direction must be positive. So, one person must be tapping in the positive y direction.

6. The position or velocity as a function of time is graphed for three objects that are traveling in a straight line. Specify which objects are experiencing a constant, nonzero net force. Explain your choices.

a.

b.

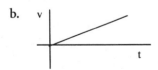

c.
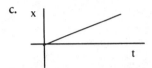

Strategy for solution: Because the force concept is involved in this problem, we should recall Newton's second law: $F = ma$. We have some kinematic information from the preceding graphs. This information can be related to Newton's second law through the acceleration. We are interested in the constant force case. Because mass is not changing for each particular object, this means we must have constant acceleration. Now we must recall what a constant acceleration says about the shape of velocity-time and position-time graphs. Remember that the acceleration tells us about the shape of the velocity-time graph.

Answers:

a. This object is not experiencing a constant net force, because the slope of the velocity graph is zero, implying zero acceleration. Because the acceleration is zero, the force on the object is zero (from Newton's second law).

b. Yes, this object is experiencing a constant, nonzero net force, because the velocity graph slope is nonzero (and constant), which implies that the acceleration is nonzero, but constant.

c. This object is not experiencing a constant, nonzero net force. The slope of the position graph is constant. This means the velocity is constant. This means that the acceleration is zero, which implies zero force.

Why is this solution successful? In each case, the written response shows how the graphical information is interpreted in terms of the important physical concepts.

7. A low-friction cart has mass 0.30 kg. A string is attached to the cart and attached to a hanging mass so that a force can be exerted on the cart. A velocity-as-a-function-of-time graph is generated for the cart.

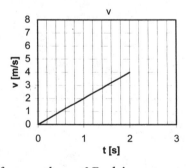

a. What is the net force on the cart? Explain your reasoning.

b. Draw the velocity graph that would result from doubling the net force acting on the cart.

> Strategy for solution: This problem combines concepts from Newton's laws of motion and kinematics. We are also dealing with interpreting a graphical representation of kinematic information. Newton's second law tells us that if we know the acceleration (and mass), then we know the force. So the problem amounts to interpreting the velocity graph appropriately to get the acceleration.

Answers:

a. Because Newton's second law tells us that $F_{net} = ma$, then we need to know the acceleration. We can get this from the velocity graph. Acceleration is the slope of the velocity graph at a particular time. Here we have a constant slope, so the acceleration is constant. Calculating the slope from the graph gives:

$$a = \frac{4.0\ \text{m}/\text{s}\text{-}0.0\ \text{m}/\text{s}}{2.0\ \text{s}\text{-}0.0\ \text{s}} = 2.0\ \text{m}/\text{s}^2$$

so

$$F_{net} = ma = (0.30\ \text{kg})(2.0\ \text{m/s}^2) = 0.60\ \text{N}$$

> Why is this a successful solution? It starts with the general principle, Newton's second law, and then explains how the important quantity, acceleration, is to be found.

b. Doubling the force will double the acceleration because $a = F_{net}/m$. So, the new acceleration is 4.0 m/s². This is the new slope for the velocity graph. So, it becomes

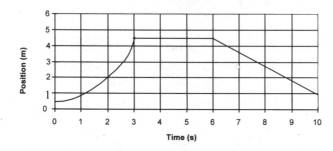

> This is a successful solution because the reasoning behind the drawing has been specified, as well as a correctly drawn graph. It is helpful that the general form of the physical law is specified (Newton's second, in this case).

8. A bowling ball rolls on a floor and collides with a golf ball that is initially at rest. While the balls are in contact with each other, how does the force of the bowling ball

on the golf ball compare with the force of the golf ball on the bowling ball? Explain your choice.

> Strategy for solution: The problem involves forces and the interaction of two objects. One place to start is Newton's laws of motion. The third law relates to interacting objects, so it is a good possibility for applying here.

Answer:

The force of the bowling ball on the golf ball has the same magnitude as the force of the golf ball on the bowling ball, but the direction is opposite. This is a direct application of Newton's third law of motion.

FINAL EXAM

> This course is not about memorizing a set of facts and formulas and reciting them back. It is important to ask of each concept developed, "How do we know that?" or "Why do we believe this?" These questions should be addressed by the students at home after each formal class period.
>
> Discussing homework assignments with colleagues is fine; however, it is best if you try each problem by yourself first. When asking for help, it is better to get hints rather than just copying the entire solution from someone.
>
> I think it is helpful to identify general principles that have been discussed that have something to do with the topic under consideration. For each general principle identified, summarize to yourself why we believe that principle. Summarize the different situations in which the principle was found to apply. Describe specific situations in your summary, and describe how the principle was used. Be clear about each step in an application. If you are unsure of any step, ask for help. Write out these summaries; don't just think about them.

1. An object is placed in front of an ultrasonic motion detector. Its motion is recorded for 10 s. The position-time graph that results is shown. Draw the corresponding velocity-time graph and acceleration-time graph. Try to be quantitative. Explain how you were able to deduce the shape of these graphs.

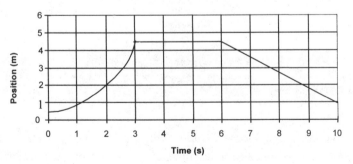

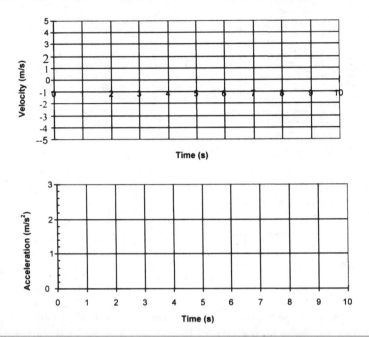

Time (s)

Strategy for solution: We need to remember some things about the graphical representation of motion data. The velocity-time graph tells us about the shape of the position-time graph. More specifically, the velocity at each time specifies the slope of the position-time graph at that particular time. Similarly, the acceleration-time graph tells us about the shape of the velocity-time graph.

The instantaneous velocity at a particular time can be estimated by calculating the average velocity over a small time interval near that time. This will allow you to make a more quantitatively correct drawing.

Answer:

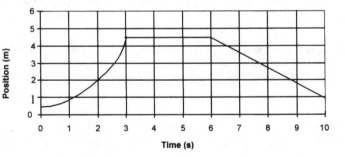

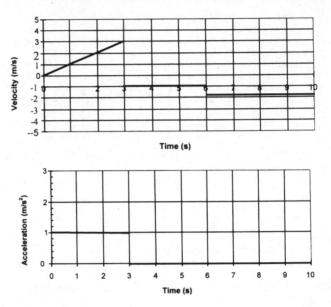

The position graph is curved from 0 to 3 s, so the velocity must be changing during this time. The velocity at $t = 0$ s is 0 m/s. The velocity near $t = 3$ s is approximately

$$\bar{v} \cong \frac{4.5 \text{ m} - 3 \text{ m}}{3 \text{ s} - 2.5 \text{ s}} = 3 \text{ m/s}$$

The position graph is horizontal from 3 to 6 s, so the velocity is zero. The position graph slopes down from 6 to 10 s, so the velocity is negative here. The slope is about –0.88 m/s.

While the velocity is changing, there must be acceleration. The slope of the velocity graph between 0 and 3 s is about 1 m/s². The velocity is horizontal elsewhere, so the acceleration is zero.

Why is this a successful response? The correct graphs have been drawn, both qualitatively correct and quantitatively correct. Some reasoning for each drawing is provided. The relationship between the velocity and position graphs and between the acceleration and velocity graphs is described.

2. A 1500-kg car sits on an icy hill, which makes a 20° angle with respect to the horizontal. The friction force between tires and road can be ignored in this situation. Arnold Schwarzenegger is pulling on the car to keep it from sliding down the hill.

   a. Draw a free-body diagram of the car, indicating all the physical forces on the car.

Strategy for solution:

1. Draw a rough picture of the situation.

2. Focus your attention on just the object of interest (the car). Draw it as a block.

3. Draw vectors for each physical force on the car.

4. Draw an xy axis system. Make the x axis run parallel to the incline. The y axis must be perpendicular to the x axis.

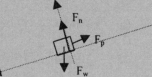

5. Make a new drawing with the force vectors coming out of the origin, using the same definition of axes as used before.

Answer:

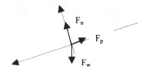

b. The car is 25 m from the bottom of the hill. Arnold releases the car. How long does it take the car to slide down the hill?

Strategy for solution: The problem involves finding a time of travel from a distance. This suggests using kinematics. This is a constant acceleration problem, so we can use our kinematic equations with constant acceleration. One of them does relate position to time. In this case, the car starts at rest at the origin, so the equation becomes

$$x = \frac{1}{2}a_x t^2$$

We need to get the acceleration. This calls for using Newton's second law. We can look at the free-body diagram for this situation to figure out the components of the forces.

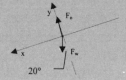

The pulling force has been eliminated, because Arnold let go of the car. Now to get the components, we need to include an angle. Inspection of the incline shows that the angle between the F_w vector and the y axis is 20°. Trigonometry tells us that the x component of the weight force is given by

$F_{wz} = mg \sin(20)$

Newton's second law, in the the x direction, will give the x component of acceleration.

Answer:

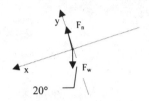

Newton's second law gives:

$$a_x = \frac{\sum F_{ix}}{m} = \frac{mg \sin(20)}{m} = g \sin(20) = 3.4 \text{ m}/\text{s}^2$$

From kinematics:

$$x = \frac{1}{2}a_x t^2 \implies t = \sqrt{\frac{2x}{a_x}} = \sqrt{\frac{2(25)}{3.4}} = 3.9 \text{ s}$$

A successful answer because the chain of reasoning is clear: It starts with the free-body diagram, progresses to applying Newton's second law, and ends with the required kinematic argument. Simply writing out the last step (the kinematic equation) would not be adequate.

3. A steel block is placed on a steel board, which is placed horizontally. Different masses are added to the block, which is then pulled with a constant velocity with a spring scale that measures the pulling force. The data obtained is shown in the following graph.

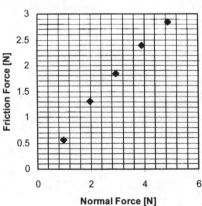

a. Why is the pulling force measured by the spring scale equal in magnitude to the kinetic friction force on the block?

Answer: The block is being pulled at constant velocity, so the acceleration must be zero. This means the total force is zero. So, the pulling force must have the same magnitude (and opposite direction) as the friction force.

b. What is the coefficient of kinetic friction for steel on steel?

Strategy for solution: The simple model of kinetic friction that we have says that the friction force is proportional to the normal force:

$$F_f = \mu_k F_n$$

This is the graph of a straight line with slope equal to the coefficient of kinetic friction. Use the graph of data points to find a best-fit straight line and find its slope.

Answer: $F_f = \mu_k F_n$, so μ_k is the slope of the friction-normal force graph. From the graph, the slope is approximately 0.62, so $\mu_k = 0.62$.

This is a successful solution because it is clear from the response what basic physics law was used to get the numerical answer. Also, it indicates the proper interpretation of the graphical information.

c. Suppose the block with total mass of 0.40 kg is traveling at 0.75 m/s when the pulling force is removed. How much work must be done on the block to stop it?

Strategy for solution: Work is an energy concept. One major energy-related law is the work-energy relationship:

$$W = \Delta KE$$

The work is done by the friction force. Here, we know the initial velocity and the final velocity (0), so the change in kinetic energy can be calculated.

Answer: $\quad W = \Delta KE = \dfrac{1}{2}mv_f^2 - \dfrac{1}{2}mv_i^2 = \dfrac{1}{2}(0.40 \text{ kg})(0.75 \text{ m}/\text{s}^2) = 0.11 \text{ J}$

4. An object of mass 0.075 kg is attached to a spring. The spring is stretched a distance of 0.15 m and released. The period of the resulting motion is measured. The measurement is repeated for a total of 20 times. The data is presented below.

0.29	
0.15	
0.21	
0.12	
0.37	
0.22	
0.20	
0.20	
0.15	
0.20	
0.22	
0.35	
0.19	
0.28	
0.30	
0.18	
0.21	
0.23	
0.19	
0.20	

a. Report the best estimate of the system's period and the uncertainty. Use the right number of significant figures.

Answer: The best estimate is the sample mean: 0.22 s. The uncertainty of the mean is given by the standard deviation of the mean = 0.014 s.

b. Now suppose the initial stretch is doubled. What is the new period of the motion? Explain why you made your choice.

Answer:

So long as the spring remains close to ideal, the period is independent of the amplitude:

$$T = 2\pi\sqrt{\frac{m}{k}}$$

So, the period is still about 0.22 s.

c. Suppose the mass is doubled, but the original stretch is used. What is the new period of the motion? Explain why you made your choice.

Answer: Assuming that the spring is ideal, then the period is given by

$$T = 2\pi\sqrt{\frac{m}{k}}$$

If m is the original mass, then the new period is given by

$$T_{new} = 2\pi\sqrt{\frac{2m}{k}} = \sqrt{2}T_{old} = \sqrt{2}(0.22 \text{ s}) = 0.31 \text{ s}$$

5. A bow is stretched with a spring scale, which can record the force required. The force readings as a function of the stretch distance from equilibrium are shown.

Stretch (m)	Force (N)
0.10	4.0
0.20	7.1
0.30	10.1
0.40	13.2
0.5	16.6

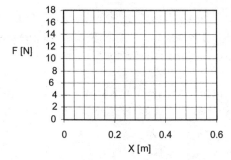

a. Does the bow obey Hooke's law? Explain your answer.

Answer: Hooke's law says that the spring force is proportional to the displacement of the spring from equilibrium. To check whether this is true here, we plot the spring force versus the spring stretch. This plot does show a proportional relationship, so Hooke's law is obeyed.

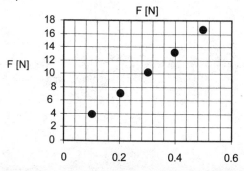

The data points are plotted on the graph so that the instructor can see the linear relationship. The relationship between the graph and the word and equation form of Hooke's law is clearly stated. These are the elements of a successful response.

b. How much work is required to stretch the bow? Explain how you arrive at your answer.

Answer: When the force on an object changes with position, then the general definition of work done by the force is used:

Work done by force = area under the force-position curve

In this case, the area can be estimated by using the triangle area formula:

$$W = \frac{1}{2}(\text{base})(\text{height}) = \frac{1}{2}(0.50 \text{ m})(16.6 \text{ N}) = 4.2 \text{ J}$$

Why is this a successful solution? In addition to writing out the equation used, this solution describes where it comes from, what general principle is being used.

c. An arrow of mass 0.023 kg and length 0.790 m is shot with this bow. The bow is stretched a distance equal to the length of the arrow. Assume that the bow obeys Hooke's law, even if it does not do so exactly. Find the speed of the arrow when it leaves the bow.

Answer: Using energy principles is a good way to approach this problem. Because the spring force is conservative, we can use conservation of energy:

$$\Delta KE = -\Delta PE$$

$$\frac{1}{2}mv_f^2 - \frac{1}{2}mv_i^2 = -\left(\frac{1}{2}kx_f^2 - \frac{1}{2}kx_i^2\right)$$

We can estimate the spring constant by finding the slope of the force-stretch graph. It is about 31 N/m. Then

$$\frac{1}{2}(0.023 \text{ kg})v_f^2 = -(0 - \frac{1}{2}(31 \text{ N}/\text{m})(0.79 \text{ m})^2)$$

$$v_f = \sqrt{\frac{31(0.79)^2}{0.23}} \text{ m}/\text{s} = 29 \text{ m}/\text{s}$$

This is a good answer because the general statement of the physical principle used is included. The origin of information not specifically mentioned in the problem statement is given (the spring constant, in this case).

CHAPTER 22

EASTERN ILLINOIS UNIVERSITY

PHY 1350: GENERAL PHYSICS I

Donald D. Pakey, Associate Professor

IVE INTERMEDIATE EXAMS AND ONE FINAL EXAM ARE GIVEN IN THIS COURSE. FOR presentation here, I have combined two of the intermediate exams as a sample midterm.

As this is usually a student's first college physics course, one of my main objectives is to teach them to be rigorous, careful, and precise in their speech, calculations, and laboratory work. I also aim to teach them the models and concepts of classical mechanics, including Newton's Laws, momentum, and energy, and show them how these can be applied in a wide range of physical situations. Another objective is for the students to enjoy the process of reading a word problem, deciding how to solve it, and then successfully doing so. Finally, I attempt to prevent the compartmentalization of their physics knowledge: they should remember the material and be able to apply it even after the course is over.

Students need to develop the following skills and competencies:

1. Conceptual understanding of the concepts and laws of classical mechanics

2. General problem-solving ability, involving the greatest amount of independent thought possible

3. Both general and specific skills in the usage of scientific instruments

4. Theory and practice of error analysis

5. Mathematical methods, most important, the theory and usage of the derivative and integral

6. Rigor and precision in thinking and problem solving

7. Understanding and solving word problems

Course examinations relate to these items as follows:

1. **Conceptual understanding of the concepts and laws of classical mechanics.** This would actually be addressed better by multiple-choice or essay-type questions, which I do sometimes use. However, it is impossible to do well on word problems, no matter how many equations the student has memorized, if there is no understanding of the concepts and when they apply.

2. **General problem-solving ability, involving the greatest amount of independent thought possible.** Because all the questions are problems, they obviously foster problem-solving ability. Sometimes, however, problems do not sufficiently encourage independent thought because the exam covers only a few chapters and the student has only to choose the correct equation from a relatively small set. Therefore it is important to have a comprehensive final exam: the wider the range of possible questions, the more closely it approximates real life.

3. **Both general and specific skills in the usage of scientific instruments.** I have only rarely included lab-related questions on exams, so this item is not addressed by the exams. The same goes for item 4, theory and practice of error analysis.

5. **Mathematical methods, most important, the theory and usage of the derivative and integral.** Several final exam problems are specifically geared toward testing mathematical facility. Problem 2 uses vector algebra, problem 3 involves simultaneous solution of equations, problem 4 involves differentiation, and problem 6 uses integration.

6. **Rigor and precision in thinking and problem solving.** This comes naturally from solving many problems on homework, quizzes, and exams. Because the students expect the exam to consist of problems, they mainly study by working problems. This goes for item 7, understanding and solving word problems, as well.

My advice to students is clearly given in the course syllabus:

How to get a good grade: Read the material and try to do the assigned problems before coming to class. I do not grade on attendance, but it's always a good idea to come to class. If you don't understand something, come talk to me (or to your classmates). To study for quizzes and exams, work out as many problems as possible, always starting "from scratch" without looking at the solution. Also, be sure to stay up to speed in your math skills—if the math gives you problems, come talk to me or to the math department tutors.

$$g = 9.80 \text{ m/s}^2 \qquad N_A = 6.022 \times 10^{23} \text{ molecules/mol}$$

Draw appropriate free-body diagrams when necessary.

1. (7 pts) Aluminum is a very lightweight metal, with a density of 2.7 g/cm³.

   a. What is the weight in pounds of a solid sphere of aluminum of radius 50 cm? (Note: A 1-kg mass corresponds to a weight of 2.2 lb. The volume of a sphere is $\frac{4}{3}\pi r^3$)

   b. If the atomic weight of aluminum is 27.0 g/mol, how many aluminum atoms are contained in this sphere?

Answers:

a. $m = \rho V = \rho \frac{4}{3}\pi r^3 = 2.7 \text{ g/cm}^3 \times \frac{4}{3}\pi (50 \text{ cm})^3 \times (1 \text{ kg}/1000 \text{ g}) = 1414 kg$

   $1414 \ kg \times (2.2 \text{ lb}/1 \text{ kg}) = 3110 \text{ lb}$

b. 6.022×10^{23} atoms/mol $\times$ (1 mol/27.0 g) $\times 1.414 \times 10^6$ g $= 3.15 \times 10^{28}$ atoms

> This problem can be done without actually knowing any equations by simply using cancellation of units. The student must not be confused by the fact that Avogadro's number is given in molecules/mol, in this case it is actually atoms/mol.

2. (8 pts) Vector A has length 4 and direction 110°, and vector B has length 6 and direction 160°.

   a. Find the components of $R = A + B$.

   $R_x =$ _____ $\qquad R_y =$ _____

   b. Find the magnitude and direction of R.

   $R =$ _____ $\qquad \theta =$ _____

Answers:

a. $R_x = A_x + B_x = 4 \cos 110° + 6 \cos 160° = -7.006$

   $R_y = A_y + B_y = 4 \sin 110° + 6 \sin 160° = 5.811$

b. $R = \text{sqrt}((-7.006)^2 + (5.811)^2) = 9.102$

   $\theta = \arccos (R_x/R) = \arccos (-7.006/9.102) = 140°$

In addition to successfully plugging the numbers into the calculator, the student should think about the answer to see if it seems about right: Do all components have the right signs? Examining the components of R shows that it is in the second quadrant, so the 140° angle is correct rather than 360° − 140° = 220°.

3. (6 pts) The pilot of an aircraft wishes to fly due west in a wind blowing at 60 km/h toward the north. If the speed of the aircraft in the absence of a wind is 225 km/h,

   a. in what direction should the aircraft head? (Describe the direction unambiguously.)

   b. what will its speed be relative to the ground?

Answers: $v_{pg} = v_{pa} + v_{ag}$, where v_{pg} is the velocity of the plane with respect to the ground, v_{pa} is the velocity of the plane with respect to the air (the direction of which the airplane's heading), and v_{ag} is the velocity of the air with respect to the ground. Because v_{pg} is required to be due west while v_{ag} is north, these three vectors form a right triangle with the hypotenuse being $v_{pa} = 225$ km/h and one side being $v_{ag} = 60$ km/h. The other side is therefore $v_{pa} = sqrt(225^2 - 60^2) = 217$ km/h, and the angle between v_{pa} and v_{pg} is arcsin (60/225) = 15.5°. Drawing the previously described right triangle shows that the plane must head in a direction of (a) 15.5° S of W, and its ground speed will be (b) 217 km/h.

In this problem, there is no substitute for remembering the formula for addition of relative velocities. The student must also know (probably from having done similar problems), or be able to figure out, that the heading of the plane is the direction of v_{pa}.

4. (9 pts) As shown, a hanging mass (10 kg) is connected by a light string over a frictionless pulley to an 8-kg box. The coefficient of kinetic friction between the box and the surface is 0.28, and the coefficient of static friction is 0.40. Find the acceleration of the masses and the tension T in the string.

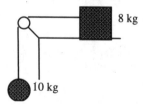

8 kg

10 kg

$a =$ _____

$T =$ _____

Answers:

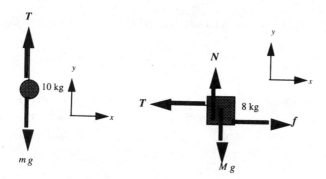

The first thing to do is to determine if the masses do in fact move. If we assume that they don't, so $a = 0$, $F = ma$ implies $T = mg = 10$ kg $\times 9.8$ m/s^2 = 98 N. Force balance in the y direction on the box gives $N = Mg = 8$ kg $\times 9.8$ m/s^2 = 78.4 N, so the maximum frictional force is $m_s N = 0.40 \times 78.4$ N = 31.4 N, so the box does in fact slide.

$F_y = ma_y$ for the hanging mass is: $T - mg = ma_y = -ma$, or $T = mg - ma$

$F_x = Ma_x$ for the hanging mass is: $f - T = Ma_x = -Ma$, or $T = f + Ma = m_k Mg + Ma$

Combining the two T equations gives $T = mg - ma = m_k Mg + Ma$, so $(m + M)\, a = mg - m_k Mg$, so $a = (m - m_k M) g / (m + M) = (10 - 0.28(8$ kg$)) 9.8$ m/s^2 /18 kg = 4.22 m/s^2

$T = m(g - a) = 10$ kg $(9.8$ m/s^2 - 4.22 m/s^2) = 55.8 m/s$^2 \bullet$ N

A proper solution must include labeled free-body diagrams, with coordinate axes, for each object. Because there are three unknowns (T, a, and f), the student should realize that he or she needs three force equations. In this case, it is obvious which three equations those are, because the x equation for the hanging mass (0 = 0) gives no information. One tricky aspect is remembering the minus signs in front of the accelerations. Alternately, the student could reverse the directions of the hanging mass y axis and the box x axis, but that nonstandard orientation of the coordinate axes is confusing to some.

Note that one could use the fact that the pulley has the sole effect of changing the direction of the tension force, so that we have a system of combined mass $m + M$ acted on by a force $mg - m_k Mg$, immediately giving a = (net force)/(total mass) = $(mg - m_k Mg)/(m + M)$. However, in this course I would give only partial credit for this solution, because I am emphasizing a systematic approach to solving $F = ma$.

5. (7 pts) A small bug is placed on a horizontal, rotating turntable at a distance 25 cm from its center. The mass of the bug is 7 g, and the coefficient of static friction between its feet and the turntable is 0.37.

   a. Find the *maximum* number of revolutions per second the turntable can have if the bug is to not slide.

b. Find the bug's speed and acceleration when it is on the verge of slipping:

$v =$ _____

$a =$ _____

Answers:

a. For a case of marginal sliding, the frictional force is equal to its maximum value $m_sN = m_smg$, so the force balance equation in the radially inward direction gives $m_smg = ma = mrw^2$. Solving for w gives $w = (m_sg/r)^{1/2} = (0.37 \bullet 9.8 \text{ m/s}^2/0.25 \text{ m})^{1/2}$ $= 3.81 \text{ rad/s} \times (1 \text{ rev}/2p \text{ rad}) = 0.606 \text{ r/s}$

b. $v = rw = 0.25 \text{ m} \bullet 3.81 \text{ rad/s} = 0.952 \text{ m/s}$

$ma = \mu_smg$, so $a = \mu_sg = 0.37 \times 9.8 \text{ m/s}^2 = 3.63 \text{ m/s}^2$

> The student must realize that this is a case of marginal sliding, so $f = \mu_sN = \mu_smg$. Then, to apply $F = ma$, he or she must realize that the relevant acceleration is the centripetal acceleration, and that revolutions per second is a measure of angular velocity. Finally, the student must have experience with using the "dimensionless dimension" radians, so that when the units of ω comes out per second (s^{-1}), he or she will know that it is actually radians per second and how to convert that to revolutions per second.

6. (6 pts) In a game of shuffleboard, a disk is given an initial speed of 8 m/s. It slides a distance of 12 m before coming to rest. What is the coefficient of kinetic friction between the disk and the surface?

Answer: Use the work-energy principle: $W = \Delta E$

$$W = -fd = \mu_kmgd = \Delta E = KE_f - KE_i = 0 - (1/2) \, m \, v^2$$

so

$$\mu_k = (1/2) \, mv^2/mgd = v^2/2gd = (8 \text{ m/s})^2/(2(9.8 \text{ m/s}^2)12 \text{ m}) = \mathbf{0.272}$$

> The student must realize that this can be solved using work-energy methods, and must then be able to write out the work and energy in terms of known quantities and solve the resulting equation. Many students run into problems when they try to calculate the work and realize that the mass is not given, because they do not see that if the mass cancels out in the end. This type of problem is useful for encouraging students to work out problems algebraically rather than numerically.
>
> It could also be done using force and acceleration, but they would still have to do it algebraically to get the mass to cancel out.

7.(9 pts) A 4-kg block is placed on a rough 15° incline, as shown. The coefficients of kinetic and static friction are 0.3 and 0.4, respectively.

a. Will the block slide down the incline?

b. What is the magnitude of the frictional force?

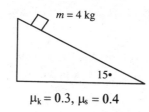

$\mu_k = 0.3$, $\mu_s = 0.4$

Answers:

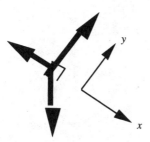

a. $F = ma = 0$ in the y direction gives $N - mg \cos 15° = 0° \rightarrow N = mg \cos 15°$. The block will slide if the x component of gravity, $mg \sin 15°$, exceeds the maximum frictional force, $m\mu_s N = \mu_s mg \cos 15°$, or equivalently, if $\sin 15° = 0.259$ exceeds $\mu_s \cos 15° = 0.386$. So, no, the block will not slide down the incline.

b. Because the block does not slide, the frictional force must be determined from force balance:

$$F = mg \sin 15° - f = ma = 0° \rightarrow f = mg \sin 15° = 4 \text{ kg } (9.8 \text{ m/s}^2) \sin 15° = 10.1 \text{ N}$$

The student must have a good understanding of the model of friction to work this problem. Even if they get part a correct, they often miss part b by assuming that the force of static friction is equal to its maximum value $\mu_s N$.

8. (8 pts) A 1.3-kg particle has a speed of 2 m/s at point A and kinetic energy of 8 J at point B. What is

a. its kinetic energy at point A?
b. its velocity at point B?
c. the total work done on the particle as it moves from A to B?

Answers:

a. $KE_a = (1/2) m v^2 = (1/2) 1.3 \text{ kg} (2 \text{ m/s})^2 = 2.6 \text{ J}$

b. $KE = (1/2) m v^2 \rightarrow v = (2KE/m)^{1/2} = (2(8 \text{ J})/1.3 \text{ kg})^{1/2} = 3.51 \text{ m/s}$

c. $W_{net} = \Delta KE = KE_B - KE_A = 8 \text{ J} - 2.6 \text{ J} = 5.4 \text{ J}$

This is basically a give-away problem to give confidence to any student who is having trouble with the rest of the exam, but it does instruct the student in the basics of work and energy.

9. (7 pts) A skydiver of mass 90 kg jumps from a slow-moving aircraft and reaches a terminal speed of 60 m/s.

a. What is the acceleration of the skydiver when her speed is 40 m/s?
   What is the drag force on the diver when her speed is:
b. 60 m/s? _____
c. 40 m/s? _____

Answers:

R

mg

a. The air drag force is proportional to the square of the speed, so it can be written $R = bv^2$, in which the constant b is determined by the condition that when $v = v_t$ (the terminal speed), $a = 0 \rightarrow R = mg$, so $b = R/v^2 = mg/v_t^2$. Therefore, the general formula is

$R = mg(v/v_t)^2$
$F = mg - R = mg[1 - (v/v_t)^2] = ma \rightarrow a = g[1 - (v/v_t)^2] =$
$9.8 \text{ m/s}^2[1 - (40/60)^2] = (5/9)g = 5.44 \text{ m/s}^2$

b. By inspection, when $v = vt$, $R = mg = 90 \text{ kg} \bullet 9.8 \text{ m/s}^2 = 882 \text{ N}$.
c. $R = mg(v/v_t)^2 = 882 \text{ N} (40/60)^2 = 392 \text{ N}$

To do this problem, the student must realize that in the formula $R = (1/2)C\rho Av^2$ we don't need to know the constants C, ρ, and A, because we know the numerical value of the terminal speed.

---

10. (9 pts) A cannon at the top edge of an 80-m cliff is fired over the edge, and the cannonball hits the ground 300 m from the base of the cliff 8 s later.
   a. With what speed was the cannonball fired? $v =$ _____
   b. At what angle above (+) or below (−) the horizontal was it fired? $q =$ _____

Answers:

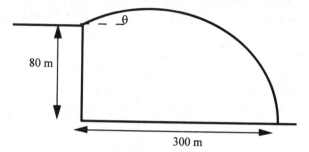

a. v_x is constant, so $v_x = \Delta x/\Delta t = 300$ m/8 s $= 37.5$ m/s.

We can find v_y from $\Delta y = v_{0y}t + (1/2)\,a_yt^2$

$v_{0y} = (\Delta y - (1/2)\,a_yt^2)/t = \Delta y/t - (1/2)(-g)t = -80$ m/8 s $+ (1/2)\,(9.8$ m/s$^2)$ 8 s $= 29.2$

$v = (37.5^2 + 29.2^2)^{1/2} = $ **47.5 m/s**

b. $\theta = \arctan(29.2/37.5) = +37.9°$

The student must remember the general method for kinematic equations, which is to find the one equation that contains all the known quantities, one unknown quantity, and nothing more. In this case it turns out to be the v_{0y} equation, which is more typically to be solved for Δy.

FINAL EXAM

For all problems, a successful response must be logically arranged and concise, with no unnecessary work. It must use the correct equations, all quantities must be correctly calculated, and proper units must be used.

For all problems, the student must show a logical progression from the data in the problem to the correct answer. The best general strategy is to read the problem, think of all equations involving the given and required quantities, choose the appropriate ones, and solve them.

For all problems, one must be able to remember equations and their meanings, quickly determine which one(s) is(are) necessary and sufficient for the solution, successfully solve the equations and do the required calculations to find the correct answer.

Show all your work and assume that all quantities are accurate to at least three significant digits.

Exam ends 12:15 P.M.

$$g = 9.80 \text{ m/s}^2 \qquad G = 6.673 \times 10^{-11} \qquad N \bullet \text{m}^2/\text{kg}^2$$
$$\text{radius of Earth} = 6370 \text{ km}$$
$$\text{density of water} = 1000 \text{ kg/m}^3$$

1. (6 pts) A 2-m³ block of wood with a mass of 1800 kg floats in water.

a. What is the weight of the water displaced by the wood?

b. What fraction of the volume of the block is above the water line?

Answers:

a. Weight of displaced water = weight of block = mg = 1800 kg • 9.80 m/s^2 = 1.76 × 10^4 N

b. Mass of displaced water = weight/g = mass of block = 1800 kg

Volume of submerged part of block = volume of displaced water = 1800 kg/1000 kg/m^3 = 1.80 m^3

Fraction of block above water line = (volume above)/(total volume) = (2.00 − 1.80)/2.00 = 0.100 (10%)

Students often have trouble expressing "fraction above water" mathematically. They must also be able to realize, or figure out, that not only the weights but the masses as well of the displaced water and the floating object are the same, according to Archimedes' Principle.

---

2. (6 pts) Vectors **A** and **B** are given by **A** = (5, −2, 3), **B** = (−1, 3, 3).

a. What is the scalar product **A** • **B**?

b. What is the angle between **A** and **B**?

c. **A** • **B** = $A_xB_x + A_yB_y = A_zB_z$ = 5 • (−1) + (−2) • 3 + 3 • 3 = −5 − 6 + 9 = −2

d. **A** • **B** = ABcos θ • cos θ = **A** • **B**/(AB) = −2[(5^2 + (−2)^2 + 3^2)((−1)^2 + 3^2 + 3)]$^{1/2}$

= −0.0744

θ = arccos(−0.0744) = **94.3°**

Nothing tricky in this problem—just a straightforward application of formulas.

---

3. (8 pts) A 5-kg mass is attached to a string, the other end of which is wound around a pulley ($I = MR^2$) of mass 6 kg and radius 0.4 m on a frictionless axle.

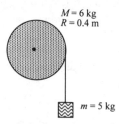

M = 6 kg
R = 0.4 m

m = 5 kg

a. When the mass is released, what is its acceleration?

b. What is the angular acceleration of the pulley?

Answers:

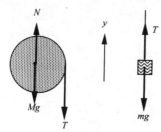

a. Clockwise torque equation for pulley, using center of pulley as origin:

$$TR = I\alpha = (1/2)MR^2\,(a/R) \Rightarrow T = (1/2)Ma$$

y force equation for mass:

$$T - mg = ma_y = -ma \Rightarrow T = mg - ma$$

Combining the two T equations:

$$T = (1/2)Ma = mg - ma \Rightarrow a(m + M/2) = mg \Rightarrow a = g/(1 + M/(2m)) =$$
$$9.80 \text{ m/s}^2/(1 + 6/(2 \bullet 5)) = 6.125 \text{ m/s}^2$$

b. $\alpha = a/R = 6.125 \text{ m/s}^2/0.4 \text{ m} = 15.3 \text{ rad/s}^2$

> The student should immediately see that this is a Newton's second law problem, but the trick is to realize that two equations are needed, one of which must be the torque equation for the pulley. The other necessary step is to recall that the relation $a = a/R$ connects a and a.

4. (8 pts) A particle starts from rest at $t = 0$ at the origin and moves in the xy plane with a constant acceleration of $a = (3i - 4j)$ m/s^2. After a time $t = 4$ s has elapsed, determine

a. the x and y components of the velocity:

$v_x =$ _____ .

$v_y =$ _____ .

b. the coordinates of the particle:

$x =$ _____ .

$y =$ _____ .

c. the speed of the particle:

$v =$ _____ .

Answers:

a. $v = v_0 + at = 0 + (3i - 4j)$ m/s² $\bullet$ 4 s =
   $12i - 16j$ m/s, so $v_x = 12$ m/s and $v_y = -16$ m/s

b. $r = r_0 + v_0 t + (1/2)at^2 =$
   $0 + 0 + (1/2)(3i - 4j)$ m/s² $\bullet$ (4 s)² = $24i - 32j$ m, so $x = 24$ m and $y = -32$ m

c. $v = (v_x^2 + v_y^2)^{1/2} = (12^2 + 16^2)^{1/2}$ m/s = 20 m/s

> The student must have learned unit vector notation and the kinematic equations in vector form.

5. (8 pts) As shown, a 4-kg block is pulled across a rough floor ($\mu = 0.3$) by a 60-N force acting at an angle of 20° above the horizontal. If the block starts from rest, how fast is it moving 4 s later?

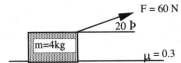

Answer:

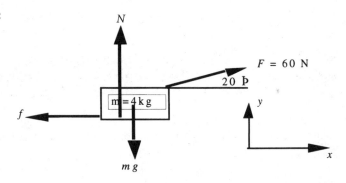

$$N + F \sin 20° - mg = ma_y = 0 \Rightarrow N = mg - F \sin 20° \Rightarrow f = mN =$$
$$\mu(mg - F \sin 20°)$$
$$ma = F \cos 20° - f = F \cos 20° - \mu(mg - F \sin 20°) \Rightarrow a =$$
$$F(\cos 20° + \mu \sin 20°)/m - mg =$$
$$60 \text{ N } (\cos 20° + 0.3 \sin 20°)/4 \text{ kg} - 0.3 \supseteq 9.80 \text{ m/s}^2 = 12.7 \text{ m/s}^2$$
$$v = v_0 + at = 0 + 12.7 \text{ m/s}^2 \; 4 \text{ s} = 50.8 \text{ m/s}$$

> None of the individual steps are particularly difficult, but there are enough steps that many students may not see immediately how to do the problem. They must break it down into pieces, first realizing that to find the speed they need to find the acceleration, and therefore they must solve Newton's second law. Many students will automatically and erroneously write $N = mg$, so this problem emphasizes that a careful solution of $F = ma$ is always necessary.

6. (7 pts) Objects of mass 3 kg and 5 kg moving in the xy plane collide and stick together. The original velocity of the 3-kg object was 10 m/s at an angle of 65°, and that of the 5-kg object was 8 m/s at 180°, both angles being measured counterclockwise from the $+x$ axis. If there are no external forces, what is the final velocity of the combined object?

magnitude _____
direction _____

Answer:
$$P_{fx} = P_{ix} = m_1 v_1 \cos \theta_1 + m_2 v_2 \cos \theta_2 =$$
3kg 10 m/s cos 65° + 5 kg 8 m/s cos 180° = –27.32 kg m/s
$$P_{fy} = P_{iy} = m_1 v_1 \sin\theta_1 + m_2 v_2 \sin \theta_2 =$$
3 kg 10 m/s sin 65° + 5 kg 8 m/s sin 180° = 27.19 kg m/s
$$P = [(-27.32)^2 + (27.19)^2]^{1/2} \text{ kg m/s} = 38.54 \text{ kg m/s}$$
$$v = P/(m_1 + m_2) = 38.54 \text{ kg m/s/8 kg} = 4.82 \text{ m/s}$$
$$\theta = \arctan (P_y/P_x) = \arctan (27.19/(-27.32)) = -44.9° + 180° = 135°$$

Other than properly using the law of conservation of momentum, the other necessary ingredient of this problem is determining that the momentum is in the second quadrant and that, therefore, the angle is 135°, not –45°.

7. (7 pts) A uniform solid cylinder ($I = MR^2$) of mass 3 kg and radius 20 cm rotates about a fixed vertical, frictionless axle with an angular speed of 8 rad/s. A 0.9-kg piece of putty is dropped vertically onto the cylinder at a point 14 cm from the axle. If the putty sticks to the cylinder, calculate the final angular speed of the system. (Assume the putty is a point particle.)

Answer:
$$L = (1/2)MR^2 \ \omega_i = [(1/2)MR^2 + mr^2] \ \omega_f \rightarrow \omega_f =$$
$$(1/2)MR^2 \ \omega_i/[(1/2)MR^2 + mr^2] = \omega_i/[1 + 2mr^2/(MR^2)] =$$
$$8 \text{ rad/s}/[1 + 2(0.9/3)(0.14/0.20)^2] = 6.18 \text{ rad/s}$$

The student must realize that angular momentum is conserved and that the moment of inertia of a point particle of mass m a distance r from the axis of rotation is mr^2.

8. (7 pts) The figure shows a thin spherical shell of mass $M = 50$ kg and radius 4 m. Two masses, 10 and 20 kg, are placed 4 m apart with the center of the shell halfway between them. What is the net gravitational force on the 10-kg mass due to the 20- and 50-kg masses?

magnitude _____
direction _____

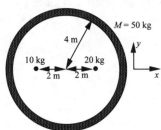

Answer:

The force due to the spherical shell is zero because the mass is inside the shell, so the entire force is due to the 20-kg mass.

$F = G \bullet m_1 \bullet m_2/d^2 = 6.673 \infty 10^{-11} N \bullet m^2/kg^2 \bullet 10 \text{ kg} \bullet 20 \text{ kg}/(4 \text{ m})^2 = 8.34 \infty 10^{-10} N$

The force is attractive, in the $+x$ direction.

The only trick here is to realize that the force due to the shell is zero.

9. (6 pts) A uniform 50-kg plank is held in place by vertical forces at A and B, as shown. What are the magnitudes and directions (up or down) of these forces?

$F_A =$ _____
direction = _____
$F_B =$ _____
direction = _____

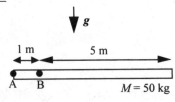

Answer:

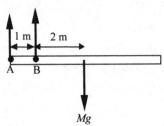

Assume that both forces are vertically upward; if a force turns out to be negative, that indicates a downward direction.

Because the plank is in equilibrium, the net torque about any point must be zero. Torques about B:

$$0 = -F_A \bullet 1 \text{ m} - Mg \bullet 2 \text{ m} \bullet F_A = -2Mg = -2 \bullet 50 \text{ kg} \bullet 9.80 \text{ m/s}^2 = -980 \text{ N (down)}$$

The net vertical force must also be zero, so

$$0 = F_A + F_B - Mg \bullet F_B = Mg - F_A = 50 \text{ kg} \bullet 9.80 \text{ m/s}^2 - (-980 \text{ N}) = 1470 \text{ N (up)}$$

> The two hints for this problem are that because the plank is uniform, Mg acts at its midpoint, and that F_A can be calculated in one step if the torque equation about B is used.

10. (7 pts) Water flows through a horizontal constricted pipe. The pressure is 3×10^5 Pa at a point at which the speed is 2.5 m/s and the area is 0.6 m². Find the speed and pressure at a point at which the area is 0.1 m².

a. $v = $ _____
b. $P = $ _____

a. Conservation of fluid → $A_1v_1 = A_2v_2$, so $v_2 = A_1v_1/A_2 = 0.6 \text{ m}^2 \supseteq 2.5 \text{ m s}/0.1 \text{ m}^2$ = 15 m/s

b. Bernoulli's equation at constant height: $P_1 + (1/2)rv_1^2 = P_2 + (1/2)rv_2^2$ → $P_2 = P_1 + (1/2)r(v_1^2 - v_2^2) = 3 \times 10^5 \text{ Pa} + (1/2) \bullet 1000 \text{ kg/m}^3((2.5 \text{ m/s})^2 - (15 \text{ m/s})^2) = 1.91 \times 10^5 \text{ Pa}$

> Students are often confused by the number of variables in Bernoulli's equation. In this problem, the clue is that because there are two unknowns we must use at least two equations, so Bernoulli's equation alone won't do the trick. The only other usable equation is the continuity equation, which immediately yields v, which can then be used with Bernoulli's equation to give P.

CHAPTER 23

GEORGIA STATE UNIVERSITY

PHYSICS 1112K: INTRODUCTORY PHYSICS

Gary Hastings, Professor

THERE ARE THREE EXAMS IN THE COURSE, INCLUDING THE FINAL. EXAMPLES OF A midterm and a final are included here.

Most students in introductory, noncalculus physics classes are fulfilling requirements usually for medical or life science–related programs. Most have forgotten rudimentary aspects of simple algebra. These issues raise concern about being in a physics class. The first course objective, therefore, is to dispel fear and any traditional myths. This can be done by showing that physics principles are at work all around us in our everyday lives.

A second objective is to generate a conceptual basis of understanding. This conceptual basis should be comprehensible, without detailed mathematical formalism.

A third objective is to develop organized reasoning and sharpen quantitative reasoning skills. These skills build on the previously mentioned conceptual understanding.

A fourth objective, not entirely related to physics, is to make students use computers.

Key competencies and skills necessary for success in this course include:

1. Recognition of key concepts and nonquantitative understanding

2. Critical thinking and reasoning

3. Ability to interpret information correctly

4. Mathematical formulation

5. Quantitative accuracy

6. Mathematical dexterity

In the exams over half the marks can be obtained purely from a strong conceptual understanding. Extra marks can be obtained if skills 2 and 3 are evident. An A grade would result if skills 1 through 4 are well developed. Little mathematical dexterity is required and most of the marks can still be obtained without a great deal of quantitative accuracy.

To emphasize conceptual understanding, in all exams, equation sheets and tables of constants can be consulted. No multiple-choice questions are given. I believe that problem-based exams are more rigorous and test all of the skills outlined previously. In addition, they test more for what a student actually understands.

Work toward gaining a strong understanding of the key concepts underlying the topics studied. These concepts will be emphasized emphatically many times during lectures. Learn how to apply them. Write them down at the start of any problem. Consult textbooks that have nonmathematical conceptual questions. Learn to be critical and try to evaluate qualitative limits on what answers could be.

Spend time really thinking about problems (without solutions in front of you) and practice, practice, practice.

MIDTERM EXAM

Closed-book exam. Answer all problems. Show all work because partial credit will be given for incomplete answers. Point value for each question is shown in parentheses. A single 8-by-11-inch sheet can be used to record all necessary formulas and constants. No other sources of information can be consulted during the exam. Duration of exam: 1 hour. Good Luck!

1. The wave shown in the figure is being sent out by a 60-cycle/s vibrator. Find the (a) amplitude, (b) frequency, (c) wavelength, (d) speed, (e) period of the wave. (8 marks)

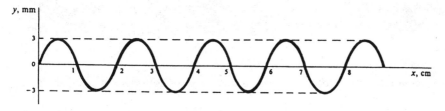

This problem deals with key concepts of waves. Parts a, b, and c are purely conceptual, but do test for ability to disseminate information from a graph. Parts d and e are quantitative, requiring the solution of simple equations. Formulation is more important than quantitative accuracy here, as most marks will be given, even if the calculation is wrong. No algebraic manipulations (dexterity) are involved in this problem.

Answers:

a. The amplitude of a wave is a measure of the maximum deviation of the wave from the undisturbed position. From the graph the amplitude is 3 mm.

b. The frequency of a wave is the number of cycles that pass a point per second. This is stated to be 60 cycles/s = 60 Hz.

c. The wavelength of a wave is the distance between any two crests or troughs or zero points. From the graph the wavelength $\lambda = 2$ cm.

d. The speed of a wave, v, is given by the equation $v = f\lambda = 60$ Hz $\times 0.02$ m $= 1.2$ m/s $= 120$ cm/s.

e. Period, $T = 1/f = 1/(60$ Hz$) = 0.0167$ s.

The response shows understanding of major points and mastery of course factual material. It also demonstrates problem-solving skills.

2. The decibel level of a jackhammer is 130 dB relative to the threshold of hearing (the threshold of hearing is 10^{-12} W/m²).

a. Determine the sound intensity produced by the jackhammer. (4 marks)

b. Determine the decibel level of two jackhammers operated side by side. (4 marks)

Here a conceptual understanding of sound intensity and intensity level is required, and of how sounds combine. Part a requires formulation of the problem, correct interpretation of data given, and some simple algebraic manipulation. Part b is almost identical and tests for conceptual understanding of how sounds combine. To emphasize conceptual awareness, in all questions, critical reasoning and conceptual knowledge are rewarded.

Extra marks can be obtained by indicating if errors have been made. For example, in part b, the answer has to be less than 260 dB but more than 130 dB. If an answer of, say, 300 or 3000 dB is obtained, and it is correctly pointed out that this is wrong and why, then extra credit will be obtained. This rationale was pointed out to students continually. If they got an answer and recognized it was wrong, and could explain why, they would get extra marks.

Answers:

a. The intensity level, dB (=130 dB), is related to the sound intensity, I, via the equation
$dB = 10 \text{ Log } (I/I_0) = 130$ dB
where

$I_0 = 10^{-12} \text{ W/m}^2$

So

$13 = \text{Log} \, (I/10^{-12} \text{ W/m}^2)$ or $(I/10^{-12} \text{ W/m}^2) = 10^{13}$

So

$I = (10^{-12} \text{ W/m}^2)(10^{13}) = 10 \text{ W/m}^2$

b. To find the decibel level, the total sound intensity must be determined.

$I_{\text{Total}} = 10 \text{ W/m}^2 + 10 \text{ W/m}^2 = 20 \text{ W/m}^2$

$dB = 10 \, \text{Log} \, (I_{\text{Total}}/I_0) = 10 \, \text{Log} \, [(20 \text{ W/m}^2)/(10^{-12} \text{ W/m}^2)]$

$\qquad\qquad = 10 \, \text{Log} \, (20 \times 10^{12} \text{ W/m}^2) = 133 \text{ dB}$

When sounds combine, the intensities can be added linearly, not the intensity levels. The answer is *not* 260 dB, but it is also greater than 130 dB.

This answer contains more information than required. It is well organized. The major points are well made, problem-solving skills are well developed. Clearly, the course material is well understood, great care is taken in presentation, and all units are included.

3. A guitar string produces 4 beats/s when sounded with a 250-Hz tuning fork and 9 beats/s when sounded with a 255-Hz tuning fork. What is the vibrational frequency of the string? (4 marks)

This problem is almost purely conceptual. It could be solved without writing any equations down. To obtain full marks, however, the reasoning process would need to be written down.

Answer: EITHER—

The beat frequency is the difference in frequency between the two sounds. In the first case, the second frequency could be 246 or 254 Hz. In the second case, the second frequency can be 246 or 264 Hz. Because the guitar string frequency is the same with both tuning forks, the frequency must be 246 Hz.

OR, MORE MATHEMATICALLY—

Beat frequency$_1 = 4 = f_{\text{fork}} - f_{\text{guitar}} \rightarrow f_{\text{guitar}} = 250 \text{ Hz} - 4 \text{ Hz} = 246 \text{ Hz}$

or

Beat frequency$_1 = 4 = f_{\text{guitar}} - f_{\text{fork}} \rightarrow f_{\text{guitar}} = 250 \text{ Hz} + 4 \text{ Hz} = 254 \text{ Hz}$

but

Beat frequency$_2 = 9 = f_{\text{fork}} - f_{\text{guitar}} \rightarrow f_{\text{guitar}} = 255 \text{ Hz} - 9 \text{ Hz} = 246 \text{ Hz}$

or

Beat frequency$_2 = 9 = f_{\text{guitar}} - f_{\text{fork}} \rightarrow f_{\text{guitar}} = 255 \text{ Hz} + 9 \text{ Hz} = 264 \text{ Hz}$

f_{guitar} is the same in both cases, so $f_{\text{guitar}} = 246 \text{ Hz}$.

Both solutions would be appropriate. Both show an understanding of major points and mastery of course factual material.

4. Two loudspeakers are separated by a distance of 10.0 m. The speakers are in phase and emit frequencies of 428.75 Hz. The speed of sound in air is 343 m/s. Point P is a position of loud sound. As one moves from P toward Q, the sound decreases in intensity.

a. How far from P will a sound minimum be heard? (6 marks)
b. How far from P will a loud sound be heard? (4 marks)

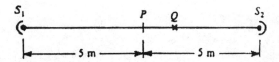

This one is extremely difficult. If the key concepts are recognized, the solution is still not straight-forward. The key concept is that the difference in the distances from the sources to Q has to be 1 or 1/2 wavelength for constructive or destructive interference, respectively. Marks would be given for stating this concept. With this in mind, one needs to calculate the wavelength. Half of the marks for the problem can be obtained by calculating the wavelength and writing down the conditions for constructive/destructive interference. The rest of the problem involves fairly difficult mathematical formulation but simple manipulation.

Again, extra marks can be obtained if the final answer is wrong and it is recognized to be incompatible with the above conditions.

Answers:

a. For destructive interference the *difference* in the distances *must* be $\lambda/2$.

b. First we need to know the wavelength of the sounds:

$\lambda = v/f = (343 \text{ m/s})/(428.75 \text{ Hz}) = 80 \text{ cm}$

Destructive interference will occur when the *difference* in the distances between the two speakers is 40 cm, and constructive interference will occur when the *difference* in the distances between the two speakers is 80 cm.

Let the distance from S_1 to P be y (so the distance from S_2 to P is also y).

Let the distance from P to Q equal x.

This means that the distance from S_1 to Q is $(y + x)$ and that the distance from S_2 to Q is $(y - x)$. The difference $S_1P - S_2P = (y + x) - (y - x) = 2x = (\lambda/2)$ for destructive interference. So:

$2x = 40 \text{ cm}$, or $x = 20 \text{ cm}$, for destructive interference

The difference $S_1P - S_2P = (y + x) - (y - x) = 2x = \lambda$ for constructive interference. So:

$2x = 80 \text{ cm}$ or $x = 40 \text{ cm}$, for constructive interference

The preceding response is complete, organized; all major points are understood and well developed; problem-solving skills are demonstrated.

5. Unpolarized light of intensity 1000 W/m passes through two polaroids. The transmission axis of the first polaroid is vertical and that of the second polaroid is at 60° to the vertical. What is the orientation and intensity of the light transmitted by the polaroids? (6 marks)

More than half the marks can be obtained from purely conceptual reasoning. The rest is simple algebra and quantitative arithmetic.

Answer: Because the light is initially unpolarized, the transmitted intensity through the first sheet is $I/2 = 500$ W/m². The light transmitted through the first polaroid has a polarization direction that is vertical. This polarized beam now makes an angle of 60° relative to the transmission axis of the second sheet. So, the intensity transmitted through the second sheet is given by Malus's Law as:

$$(500 \text{ W/m}^2) \cos^2 60° = (500 \text{ W/m}^2)(0.25) = 125 \text{ W/m}^2$$

The response is complete, organized; all major points are understood and well developed; problem-solving skills are demonstrated.

6. A 1.50-cm-high diamond ring is placed 20.0 cm from a concave mirror whose radius of curvature is 30.0 cm.
   a. Draw a ray diagram to locate (approximately) the position and orientation of the image. (4 marks)
   b. Determine quantitatively the position of the image of the diamond ring. (4 marks)
   c. Determine quantitatively the size of the image. (4 marks)

Both qualitative and quantitative reasoning are required here. Ray diagrams are not requested but would be useful for recognizing incompatibilities between the solutions to equations.

To solve this problem, one needs to be very well versed with sign conventions for mirrors and be able to manipulate fairly complicated equations. First, one needs to recognize that the radius of curvature of a mirror is related to the focal length. By calculating the focal length, it can then be incorporated into another equation and the image distance can be obtained. This image distance should be compared with the ray diagram. Great care has to be taken in calculating the image distance because the equation is complicated. The results of all algebraic manipulations should be compatible with ray diagrams.

Answers: I would draw a ray diagram to start with. From this I would clearly see that the image is inverted and magnified and farther from the mirror than the object. The quantitative equations should all agree with this:

The focal length, $f = R/2 = (30.0 \text{ cm})/2 = +15.0 \text{ cm}$

$d_0 = +20 \text{ cm}, \quad f = +15 \text{ cm}, \quad d_i = ?$

$d_i = (f \bullet d_0)/(d_0 - f) = (15 \times 20 \text{ cm}^2)/(+5 \text{ cm}) = +60 \text{ cm}$

The positive sign indicates that the image is to the left of the mirror, and is therefore a real image. This should agree with the ray diagram.

Magnification, $m = - d_i d_0 = -(+60 \text{ cm})/(+20 \text{ cm}) = -3$

The image is magnified because the magnification is greater than one, and inverted because of the negative sign.

Image height, $h_i = m h_0 = -3(1.5 \text{ cm}) = -4.5 \text{ cm}$

The minus sign indicates that the image is inverted.

> The response is complete and organized. It also contains critical evaluation of results and an understanding of major points, demonstration of problem-solving skills, and demonstration of mastery of course factual material.

FINAL EXAM

Closed-book exam. Answer all problems and show all work because partial credit will be given for incomplete answers. Point value for each question is shown in parentheses. Two 8-by-11-inch sheets can be used to record all necessary formulas and constants. No other source of information can be consulted during exam. Duration of exam: 2 hours. Good Luck.

1. A 1.50-m string is stretched so that the tension is 2.4 N. A transverse wave with a frequency of 120 Hz and a wavelength of 0.5 m travels along the string.

   a. What is the speed of the wave along the string? (3 marks)
   b. What is the mass of the string? (4 marks)

> Part a requires an understanding that frequency and wavelength are related to wave velocity. The calculation is simple, and virtually all students successfully completed this part. Part b uses the results of part a, and some manipulation to obtain the mass.

Answers:

a. The frequency and wavelength of the wave are given, so the wave speed can be calculated:

$v = f\lambda = 120$ Hz $\times$ 0.5m $= 60$ m/s

b. The wave speed is related to the linear density (and hence mass) through the relation:

$$v = \sqrt{\frac{F}{m/l}} \quad \text{So} \quad v^2(m/l) = F \quad \text{or} \quad m = \frac{Fl}{v^2}$$

$$m = \frac{2.4\,N \times 1.5\,\text{m}}{3600(\text{m}/\text{s})^2} = 10^{-3} \text{ kg} = 1 \text{ g}$$

The response is complete. It is an understanding of major points and course material. Problem-solving skills are demonstrated.

2. The intensity level of a sound produced when a gun is fired is 110 dB, relative to the threshold of hearing.

   a. Determine the sound intensity produced by the gun. (5 marks)
   b. Determine the decibel level of two identical guns fired simultaneously. (4 marks)

   Assume that the threshold of hearing is 10^{-12} W/m^2.

This is almost identical to problem 2 of the midterm exam.

Answers:

a. The intensity level, dB ($=110$ dB), is related to the sound intensity, I, via the equation

$dB = 10 \text{ Log } (I/I_0) = 130$ dB
where
$I_0 = 10^{-12}$ W/m^2

So

$11 = \text{Log } (I/10^{-12} \text{ W/m}^2)$
or
$(I/10^{-12} \text{ W/m}^2) = 10^{11}$
so
$I = (10^{-12} \text{ W/m}^2)(10^{11}) = 0.1$ W/m^2

When sounds combine, the intensities can be added linearly, not the intensity levels. The answer is *not* 220 dB, but it is also greater than 110 dB.

b. To find the decibel level, the total sound intensity must be determined:

$I_{Total} = 0.10 \text{ W/m}^2 + 0.10 \text{ W/m}^2 = 0.20 \text{ W/m}^2$

The total intensity level is:

$dB = 10 \text{ Log } (I_{Total}/I_0) = 10 \text{ Log } [(0.2 \text{ W/m}^2)/(10^{-12} \text{ W/m}^2)]$
$= 10 \bullet \text{Log}(0.20 \times 10^{12} \text{ W/m}^2) = 113 \text{ dB}$

The answer is almost the same as for problem 2 of the test. It contains more information than required. It is well organized. The major points are well understood, and problem-solving skills are well developed. Algebraic skills are also demonstrated. Clearly, the course material is well understood, great care is taken in presentation, and all units are included.

3. The following figure shows a ray of light that originates in an aquarium. It travels through the water and is incident on the glass at an incident angle of 45°. Part of the light is reflected, and part of the light is refracted. The refracted portion then emerges into the air. *Note:* Ignore any partial reflections at the glass/air interface. Find the following:

a. The angle of reflection, θ. (2 marks)
b. The angle of refraction, ϕ. (3 marks)
c. The angle between the reflected and the refracted rays, α. (3 marks)
d. The second angle of refraction, β. (4 marks)

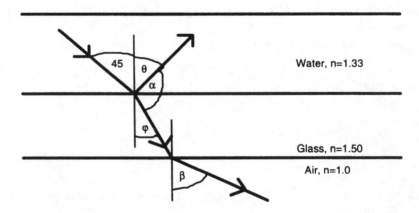

This item tests for an understanding of the law of reflection and refraction. Part a simply requires one to write down the law of reflection. Parts b and d are simply exercises in the application of Snell's law. It should be obvious that the refraction angle, ϕ, is less than 45° and that the refraction angle, β, is greater than 45°. Again, it is possible to gain marks by indicating in what range the angles should lie, based on purely geometric grounds, or from a simple understanding of refraction at an interface between two media. Part c is purely an exercise in geometry.

Answers:

a. The law of reflection states that the angle of incidence equals the angle of reflection. So, from the diagram, the angle of reflection is 45°.

b. Snell's law states that:

$$n_1 \sin \theta_1 = n_2 \sin \theta_2$$

where n_1/n_2 is the refractive index of the first/second medium.

Also note that in the diagram, $\theta > \phi$, and that the ray bends toward the normal when entering a denser medium (higher refractive index). So $1.33 \sin 45 = 1.5 \sin \phi_2$. So $\phi = 38.83°$.

c. From simple geometry, the angle can be obtained:

$$\alpha = 180 - \theta - \phi = 180 - 45 - 38.33 = 96.17°$$

d. Again, from Snell's law:

$$1.5 \sin \phi = 1.0 \sin \beta$$

Note that in the diagram, $\beta > \phi$, and that the ray bends away from the normal when entering a less dense medium (lower refractive index). So, $1.5 \sin 38.33 = 1.0 \sin \beta_2$. So, $\beta = 70.13°$.

The answer contains more information than required. It is well organized. All major points are well understood. Mastery of course material is obvious, and problem-solving skills are well developed. Algebraic skills are also demonstrated.

4. A *vertically polarized* beam of light is incident on a group of three polarizing sheets (see figure). The transmission axis of each sheet is rotated 45° with respect to the preceding sheet. The transmission axis of the first sheet is vertical (see figure).

   a. What fraction of the incident light intensity is transmitted? (6 marks)
   b. If the second sheet is removed, what fraction of the incident light intensity is transmitted? (2 marks)

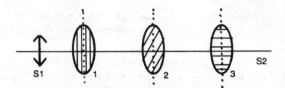

This item is very similar in scope to the next problem. Part b is a purely conceptual exercise requiring recognition that crossed polarizers transmit no light.

Answers:

a. The transmission axis of the first polarizer is parallel to the polarization of the incident light. So *all* the light passes through the first polarizer. Let the incident light intensity be I. The light transmitted through the first polaroid has a polarization direction that is vertical.

This polarized beam now makes an angle of 45°, relative to the transmission axis of the second sheet. So, the intensity transmitted through the second sheet is given by Malus's law as:

$$(I) \cos^2 45° = I/2$$

The light transmitted through the second polaroid has a polarization direction that is at 45°, relative to the vertical direction. This polarized beam now makes an angle of 45°, relative to the transmission axis of the third sheet. So, the intensity transmitted through the third sheet is given by Malus's law as:

$$(I/2) \cos^2 45° = I/4$$

Only 25 percent of the incident light is transmitted by the three polarizing sheets.

b. If the second sheet is removed, the first and third polarizing sheets are *crossed* (at 90° to each other), and zero light is transmitted.

The response is complete and organized. All major points are understood and well developed, problem-solving skills are demonstrated.

5. A 2-cm-high object is placed 30 cm in front of a *concave* spherical mirror, which has a radius of curvature of 40 cm.

a. What is the focal length of the mirror? (2 marks)
b. Where is the image located relative to the mirror? (4 marks)
c. Is the image real or virtual? Explain why. (2 marks)
d. What is the magnification? (3 marks)

e. What is the height of the image? (3 marks)

f. Is the image upright or inverted? Explain why. (2 marks)

> This item involves application of thin-lens equations. If all the rules are understood, then the problem is easy. In my experience, errors are frequent, and physically unreasonable answers result. Most students are usually unaware of this. In this question, I should have insisted upon drawing a ray diagram, because this is important for checking the results of calculations. At least, if a discrepancy is noted, extra credit could then be given.

Answers: I would draw a ray diagram to start with. From this it is clear that the image is inverted, magnified, and further from the mirror than the object. The quantitative equations should all agree with this:

a. The focal length, $f = R/2 = (40.0 \text{ cm})/2 = +20.0 \text{ cm}$

b. $d_0 = +30 \text{ cm}$, $f = +20 \text{ cm}$, $d_i = ?$

$d_i = (f \bullet d_0)/(d_0 - f) = (20 \times 30 \text{ cm}^2)/(+10 \text{ cm}) = +60 \text{ cm}$

The positive sign indicates that the image is to the left of the mirror, and is therefore a real image. This should agree with the ray diagram.

c. See the preceding: the image is real.

d. Magnification, $m = -d_{-i}/d_0 = -(+60 \text{ cm})/(+30 \text{ cm}) = -2$

The image is magnified because the magnification is greater than one, and inverted because of the negative sign.

e. Image height, $h_i = mh_0 = -2(2.0 \text{ cm}) = -4.0 \text{ cm}$

f. The minus sign indicates that the image is inverted.

> The response is complete and organized. It also contains critical evaluation of results and an understanding of major points, demonstration of problem-solving skills, and a demonstration of mastery of course factual material.

6. A diverging lens has a focal length of −10 cm. An object is placed 30 cm from the lens.

a. Determine the distance between the object and the image. (4 marks)

b. Draw a ray diagram showing the position and orientation of the image. (3 marks)

c. Is the image real or virtual? (1 mark)

d. Is the image upright or inverted? (1 mark)

e. Is the image reduced or magnified? (1 mark)

A straightforward application of lens equations. Parts c and d are multiple choice and can be answered without any equations, simply by drawing ray diagrams. I frequently encourage my students to draw diagrams and do the math and cross-check the two.

Answers: Again, I would draw a ray diagram to start with. From this it is clear that the image is upright, virtual, and reduced in size. It is also closer to the lens than the object. Its distance from the lens is actually less than the focal length of the lens. The quantitative equations should all agree with this:

a. $d_0 = +30$ cm, $f = -10$ cm, $d_1 = ?$

$d_i = (f \bullet d_0)/(d_0 - f) = (-10 \times 30$ cm$^2)/(+40$ cm$) = -7.5$ cm

The negative sign indicates that the image is to the left of the lens, and is therefore a virtual image. This should agree with the ray diagram. The object-to-image distance is 30 cm − 7.5 cm = 22.5 cm.

b. Magnification, $m = -d_i/d_0 = -(-7.50$ cm$)/(+30$ cm$) = +0.25$

The image is reduced in size because the magnification is less than one, and upright because of the positive sign.

c. The image is virtual because it is to the left of the lens.

d. The image is upright because the magnification is positive.

The response is complete and organized. It also contains critical evaluation of results, an understanding of major points, demonstration of problem-solving skills, and demonstration of mastery of course factual material.

7. A spaceship traveling east flies directly over the head of an inertial observer who is at rest on the earth's surface. The speed of the spaceship is such that

$$\sqrt{1 - \frac{v^2}{c^2}} = \frac{1}{2}$$

a. The navigator on the spaceship observes a neon sign on a storefront. If he measures the speed of light emitted from the sign *as he approaches* the sign, what value will he obtain? (2 marks)

b. The navigator's onboard instruments indicate that the length of the ship is 20 m. If the length of the ship is measured by an earthbound inertial observer, what value will be obtained? (3 marks)

c. The pilot fires an ion gun, which propels ions from the ship at 1.0×10^8 m/s, relative to the ship, toward the earth. What is the speed of the ions as measured by the earth observer? (6 marks)

d. An apple falls from a tree and the earth observer measures that it takes 4 s to reach the ground. According to the navigator on the ship, how long does the apple take to reach the ground? (3 marks)

Part a requires knowledge that the speed of light is the same in any reference frame, which is, in fact, one of the fundamental postulates of special relativity. Nearly every student in the class recognized this.

Part b requires one to understand that moving objects appear contracted. If a number greater than 20 m was calculated and a student recognized the invalidity of this number, extra credit would be obtained.

Part d is similar, so long as it is recognized that time appears to dilate for a moving observer. Most students successfully completed parts a, b, and d.

Part c is very difficult. First, the velocity of the ship needs to be calculated from the equation given. Then, an equation for the relative velocities is written down. It is still difficult to assign the variables in the equation correctly. With hindsight, I would not include part c in future exams. Virtually none of my students had any intuitive understanding of part c. (In the future, I shall not cover special relativity. The only reason, as far as I can tell, for including special relativity is to obtain the equation $E = mc^2$, which is used a great deal in the study of quantum mechanics, radioactivity, and nuclear physics.)

Answers:

a. The speed of light is the same in any frame of reference. Therefore, the navigator measures a speed of 3×10^8 m/s for the light from the sign.

b. The navigator on the ship measures the proper length, $L_0 = 20$. Moving objects appear to contract, so the earthbound observer measures a contracted length for the ship, $L = \gamma/L_0$, where $\gamma = 2$, is given in the problem. So, $L = 2/20$ m $= 10$ m. Length is contracted.

c. To the earth-based observer, the rocket ship approaches at speed V, and the ion beam at speed U', relative to the ship. The speed of the ion beam relative to the earth observer is U. So:

$U' = 1 \times 10^8$ m/s, $V = ?$ (need to calculate)

Now from the given equation

$[1 - (V^2)/c^2] = 0.25$

so

$V^2/c^2 = 0.75$

or

$V = 0.866 \times c = 2.6 \times 10^8$ m/s

so

$U = (U' + V)/[1 + (U'V)/c^2]$
$= (3.6 \times 10^8 \text{ m/s})/[1 + (1 \times 10^8 \text{ m/s})(2.6 \times 10^8 \text{ m/s})/c^2]$
$= (3.6 \times 10^8 \text{ m/s})/[1 + (2.6 \times 10^{16})/(3 \times 10^{16})]$
$= (3.6 \times 10^8 \text{ m/s})/[1 + 1.29]$

$$= 2.8 \times 10^8 \text{ m/s}$$

d. The earth observer measures the proper time, $t_0 = 4$ s. Moving objects appear to take longer so, from the rocket ship, the apple appears to take longer to hit the ground. The time measured by the observer on the ship, $t = \gamma \times t_0$, where $\gamma = 2$, is given in the problem. So, $t = 2 \times 4$ s $= 8$ s. Time is dilated.

> The response is complete and organized. It clearly shows mastery of course factual material and an understanding of all major points. Problem-solving skills are well developed, and ability to manipulate algebraic equations is demonstrated.

8. A material has a photoelectric work function of 3.66 eV. What is the maximum velocity of the photoelectrons ejected from it by light of wavelength of 250 nm? (12 marks)

> This problem requires knowledge of several important points. The key point is to recognize that photons carry energy, and all this energy is transferred to the metal and the electron with the subsequent destruction of the photon (conservation of energy). The wavelength is given, so the photon energy can be obtained. The work function is given, so the energy transferred to the metal is known. But, it is given in electron volts. A conversion to joules is required (because Planck's constant has units of joule seconds). From conservation of energy, the kinetic energy of the electron can be obtained, and from this the velocity of the electron is obtained.
>
> This problem can be solved in two or three stages of number crunching. Or two equations can be written down, and a new equation for the electron velocity can be derived. Most of my students used the former approach. I thought this was a difficult problem requiring much organized thought and foresight; however, most of the students in my class managed to complete it successfully.

Answer: The surface has a work function,

$$W_0 = 3.66 \text{ eV} = 1.6 \times 10^{19} \text{ J} \times 3.66 \text{ eV} = 5.85 \times 10^{-19} \text{ J}$$

The wavelength of the light is given, so the energy of the photon can be calculated:

$$E = hf = hc/\lambda = (6.626 \times 10^{-34} \text{ Js})(3 \times 10^8 \text{ m/s}) /(250 \times 10^9 \text{ m}) = 7.92 \times 10^{-19} \text{ J}$$

Energy is conserved in the process, and the photon energy is transferred to the electron and the material. That is to say:

$$hc/\lambda = W_0 + KE$$

or

$$KE = 7.92 \times 10^{-19} \text{ J} - 5.85 \times 10^{-19} \text{ J} = 2.064 \times 10^{-19} \text{ J}$$

But

$$KE = 1/2mv^2$$

or

$$v^2 = (2 \times KE)/m$$

$v^2 = (4.128 \times 10^{-19} \text{ J}) / (9.1 \times 10^{-31} \text{ kg}) = 4.536 \times 10^{11} \text{ (m/s)}^2$

So

$v = 6.74 \times 10^5 \text{ m/s}$

> Facts are well understood. Problem-solving skills are well developed, and ability to manipulate algebraic equations is demonstrated. The response is complete and organized.

9. Find the energy (in joules) of the photon that is emitted when the electron in the hydrogen atom undergoes a transition from the $n = 7$ energy level to the $n = 3$ energy level.

> Calls for knowledge of photon emission upon electron transitions between different energy states. The key concept, again, is conservation of energy. The photon has energy that is equal to the difference in the energies of the two states. The energies of the different states can easily be written down and the difference calculated. It should be noted that the problem can also be solved by calculating the wavelength from the equations that govern the Paschen series, and then solving for the energy.

Answer: When an electron makes a transition between energy levels, a photon is emitted with an energy equal to the difference in energies of the two levels. For the seventh energy level, the energy is $E_7 = -(13.6 \text{ eV})/7^2 = -0.278 \text{ eV}$. For the third energy level, the energy is $E_3 = -(13.6 \text{ eV})/3^2 = -1.51 \text{ eV}$. The difference in energy of the two levels is $E_3 - E_7 = 1.23 \text{ eV} = 1.97 \times 10^{-19} \text{ J}$.

> Factual material is well understood, as are all major points.

10. An electron in a hydrogen atom is in the $1 = 3$ subshell (the orbital angular momentum quantum number is $1 = 3$).

   a. How many states are in this subshell? (3 marks)
   b. What is the minimum value for the total energy of this electron? Give your answer in electron volts. (5 marks)

> Problem 10 expands upon problem 9 by testing for knowledge of the substructure of energy states. Students have to realize that electrons can exist in finite energy states, and that each state has a number of substates but all with the same total energy.

Answer: In the $1 = 3$ subshell, there are $2(21 + 1)$ substates $= 2(6 + 1) = 14$. The total = energy quantum number, n, is given by the equation:

$n > 1 + 1$ or $n = 1 + 1$

The minimum value of n is $3 + 1 = 4$. Therefore, the minimum value of the energy of the electron is:

$E_4 = -(13.6 \text{ eV})/4^2 = -0.85 \text{ eV}$

If the quantum number, n, were larger, the total energy would be larger.

> Factual material is well understood, as are all major points.
>
> It should be clear from the responses in these exams that a good answer contains text describing relevant concepts and how they are formulated in terms of algebraic equations. Critical reasoning is included by analyzing what the possible answers could or could not be. Also, the overall thought process is organized, and clear use and understanding of units is demonstrated throughout.

CHAPTER 24

UNIVERSITY OF ILLINOIS

PHYSICS 101: INTRODUCTORY PHYSICS

Ian K. Robinson, Professor

PHYSICS 101 IS THE FIRST SEMESTER OF A TWO-SEMESTER INTRODUCTORY NONCAL-culus physics course. The exams in this course test three major skills and competencies:

1. Your ability to express an observation in the real world as an equation or equations.

2. Your ability to recognize in which everyday situations the outcome can be predicted by the laws of physics.

3. Your ability to appreciate the logic behind the history of science.

In addition, Professor Robinson says he looks for a demonstration of your recognition of which quantities in the description of the problem correspond to which symbols in the equations. Also important are an ability to construct abstract diagrams connecting vectorial quantities and competence with basic algebraic manipulation.

Perhaps the most important thing you can do to prepare for exams in this course is to practice problem solving. In addition, be sure to observe the following:

◆ Do at least a quick reading on the topics to be covered in the lecture *before* attending class. When you read prior to a lecture, do so actively rather than passively. Try to identify a few questions or key points to listen for.

◆ Be alert during the lecture. Most instructors emphasize and highlight the most important points.

◆ Take good notes and review them.

◆ In all science courses, particularly those that rely heavily on mathematics and problem solving, it is extremely important that you do the homework assignments regularly and keep up with the class. Do not let yourself fall behind.

It is a rare physics exam that presents you with all new material. Although there may well be a few surprises, most of the problems presented should be familiar to you. Some may even be drawn directly from lab sessions. Often, problems presented in lecture will appear on the exam, perhaps with the numbers changed. Take careful notes of all problems presented in class.

Even more common is the reappearance of homework problems on exams. This is one reason why it is so important to keep up with homework assignments. It might also be helpful to form a study group to work homework problems together; however, it is usually best to do your final studying alone.

Know the math. A basic repertoire of equations and manipulations of logarithms and exponents is common to most physics problems. Practice any mathematical operations with which you are unfamiliar. Learn the major laws, formulas, and equations. The formula sheets provided with the exams for this course are a great help, but you need to be familiar with the tools to use them effectively. In some cases, solving a problem will at least begin with plugging numbers into equations obtained from the formula sheet. Of course, you need to know when to do this. But, you can be equally certain that at least one part of each problem will require algebraic manipulation of more than one standard equation.

Professor Robinson points out that the exams are graded manually, and liberal partial credit is given. The multiple-part problems are usually sequential. If you can see how to get, for example, from the answer to part a to part b, you will get credit even if you make up or guess the answer to part a.

The exams in the course are complex. It is not expected that students will be able to answer all of the questions in the 60 minutes allotted. We aim for about 70 percent. Look at the points distribution, Professor Robinson advises. In a 60-minute test, you should spend no more than 15 minutes on the multiple-choice parts. These questions will, therefore, not be too difficult. If you get stuck, move on!

FIRST HOUR EXAM

A few tips that will help you once you sit down to take the exam:

• Don't just dig in! Take a few moments to look over all of the questions. Identify those you feel most comfortable with and do them first. Save the most novel or difficult for last. Remember, the exams for this course are not designed to be completed in the time allotted. Make certain you address all the problems you feel confident of.

• Note the point distribution of the questions. Gauge your time accordingly.

• Don't forget to use the formula sheet!

• Remember to include the appropriate units in your answer (kilograms, miles per hour, whatever). Omitting units is a careless error.

• In showing your work, write clearly and legibly. Your instructor cannot assign partial credit for a wrong answer if he cannot see what you have done. Many wrong answers are the result of computational errors. To a significant degree, these will be forgiven *if* it is clear that the underlying principles of a problem are understood.

• Take a common-sense look at your answer. Does it make sense? Is it plausible? If not, you may have made a simple (but fatal) error, such as overlooking a minus sign on a negative number.

Part 1—Multiple choice. *Circle the correct answer. If a calculation is required you* must *show your work to receive credit. There is no partial credit in part 1.*

M1. (7 pts) A rescue plane flying at a speed of 120 km/h sets off to meet up with a ship located 100 km due east at the time of departure. The ship is cruising north at 50 km/h. What course should the pilot steer?

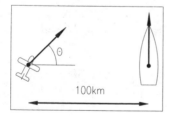

100km

a. $\theta = 22.6°$
b. $\theta = 24.6°$
c. $\theta = 65.4°$
d. $\theta = 67.4°$

Work shown:

$$\sin^{-1}\frac{50}{120}$$

Answer: b

This is a vector addition problem. First, draw triangle of velocities; next, identify sides; finally, use trigonometry. You must understand the graphical meaning of vectors to answer the question.

M2. (7 pts) A block of mass 2 kg is resting on a 20°sloping plane. The coefficient of static friction between the surfaces is $\mu_S = 0.55$, while the coefficient of kinetic friction is $\mu_K = 0.35$. What is the friction force holding the block at rest?

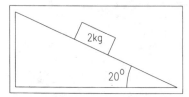

a. $f = 3.7$ N
b. $f = 6.4$ N
c. $f = 6.7$ N
d. $f = 8.3$ N
e. $f = 10.1$ N

Work shown:

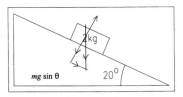

$f = mg \sin g$

Answer: c

Begin by drawing a free-body diagram with f, N, and W; set $\Sigma F = 0$ in x and y directions; then, solve for f. You must locate all the forces and use trig to identify their direction.

M3. (7 pts) The position of a deranged rat as it runs along a sewer pipe is plotted as a function of time in the picture. What is its average velocity between points A and D?

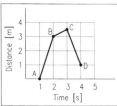

a. 0.25 m/s
b. 0.33 m/s
c. 3.5 m/s
d. 9 m/s

Answer: b

This question requires the ability to read a graph. To answer the question, read off points A and D, then determine Δx and Δt. Next, calculate $\bar{v}$ from the formula sheet. It must be understood that the concept of "average" does not depend on intermediate motion.

M4. (7 pts) Two Atwood's machines A and B are constructed from light, frictionless pulleys, light nonextending ropes, and hanging masses. Machine A has a 10-kg mass on the right side, while machine B is pulled by hand with a force of 98 N. Which machine has the greater acceleration?

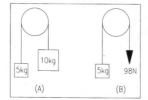

a. A greater
b. B greater
c. both the same

Answer: b

The applicable formula is:

$$a = \left(\frac{m_2 - m_1}{m_2 + m_1} \right) g$$

and the problem is to determine the difference between mass and weight. Begin with a free-body diagram for the 10-kg mass.

$$\Sigma f_y = ma$$

is not zero. Determine the tension of rope A and compare with B. Some students complained that this was a trick question. Actually, it simply shows that the tension is not equal to mg when an object accelerates.

Part 2—Problems. *Show your work. Partial credit will be given if it is clear from your work that you are making progress.*

P1. (24 pts) A skier skis horizontally over the edge of an 8-m cliff with a velocity of 15 m/s as shown in the diagram.

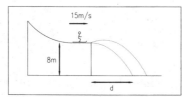

In this trajectory problem, understanding the independence of motion in the x and y directions is critical. The equations are supplied on a formula sheet, but the student must identify the symbols with the information given.

a. How long is the skier in the air?

Work shown:

$$y = y_0 + v_{oy't} - \frac{1}{2}gt^2$$

$$t = \sqrt{\frac{2(y_0 - y)}{g}}$$

$$(0)(8 \text{ m})$$

Answer:

$t = 1.28$ s

Note that "horizontal" means $v_y = 0$; apply the uniform acceleration equation to motion in y; then solve for Δt.

b. How far away from the cliff does the skier land?

Work shown:

$$x = x_0 + v_{ox}t$$

$$(0)(15 \text{ m/s})$$

Answer:

$d = 19.2$ m

See that Δt in x is the same as Δt in y. The horizontal velocity does not change. Use the kinematic equation with v_x and Δt.

c. What is the skier's speed of impact with the snow?

Work shown:

$$v_y = v_{oy} - gt = -12.5 \text{ m/s}$$

$$|v| = \sqrt{(12.5)^2 + (15)^2}$$

Answer: $|v| = 19.5$ m/s

1. Speed = magnitude of velocity (Pythagorean triangle of v_x and v_y).
2. Find v_y from the kinematic equation.
3. Combine v_x and v_y.

d. The skier finds it is much more fun to jump upward just as he or she is leaving the cliff. With some practice, the skier finds he or she can take off at an angle of

15° with the same 15-m/s velocity (magnitude). How much higher than the edge of the cliff does the skier fly?

Work shown:

$$v_y^2 = (v_0 \sin 0)^2 - 2g\Delta y$$

v_y^2 is 0 at top

$$\Delta y = \frac{(v_0 \sin 0)^2}{2g}$$

Answer:

$h = 0.77$ m

> Realize that the motion is no longer horizontal, then find v_y using trigonometry. Finally, use the kinematic equation (time-independent version) to find distance traveled in y.

P2. (24 pts) A stack of books, each of weight 15 N, is lying on a table without moving.

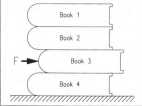

> The phrase "without moving" tells us this is a static equilibrium problem. Students are expected to find the normal forces (both up and down) at each book-book contact point. The solution also uses Newton's third law. In solving word problems, careful reading is an essential skill. In a well-constructed word problem, every word is significant.

a. What is the normal force acting from Book 2 on Book 3?

Work shown: 2×15 N

Answer:

$F_{23} = 30$ N

> 1. Draw a free-body diagram for each book.
> 2. Connect the forces between the books, using Newton's third law.
> 3. Find the required force by systematic algebraic reduction.

b. What is the normal force acting from Book 4 on Book 3?

> This answer follows the form in part a.

Work shown:

$3 \times 15\,\text{N}$

Answer:

$F_{43} = 45\,\text{N}$

c. A sideways force *F* is applied to Book 3, as shown. Draw a free-body diagram for Book 3, indicating all the forces acting on it.

Answer:

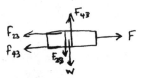

Draw the diagram, including force *F* and friction forces that oppose it. Label the magnitudes.

d. If the coefficient of friction between the books is $\mu_S = 0.3$, how much sideways force *F* must be applied to Book 3 to start it moving, assuming the books above it are constrained (sideways) not to move with it?

Work shown:

$$\left.\begin{array}{l} f_{23} = \mu_s F_{23} \\ f_{43} = \mu_s F_{43} \end{array}\right\} 0.3 \times 75\,\text{N}$$

Answer:

$F = 22.5\,\text{N}$

Knowing μ_s and normal forces allows us to find *f*. *Hint:* There are two friction forces! Note that some students are uncomfortable with the idea that normal forces can point down as well as up.

P3. (24 pts) A block of mass 3 kg is connected via a light string, over an ideal pulley to a hanging mass of 0.5 kg. The table is flat and frictionless.

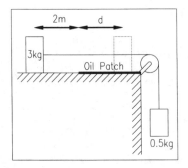

Work shown for all questions:

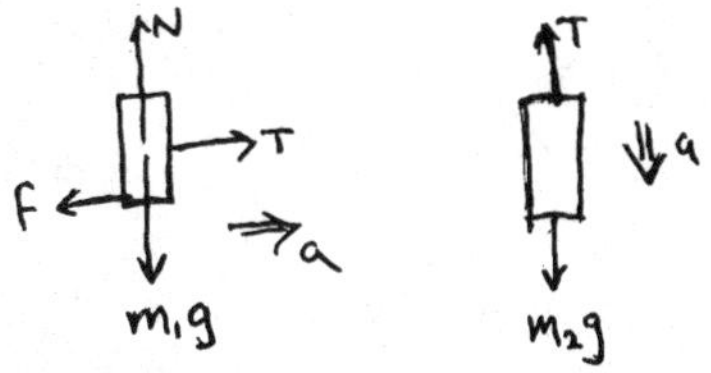

For this multibody dynamics problem with constrained motions (and kinetic friction), begin by drawing free-body diagrams for objects 1 (3 kg) and 2 (0.5 kg). Note that $\sum F = ma$ for objects 1 and 2 and that coupling means that a_x(3 kg) $= -a_y$(0.5 kg).

a. What is the initial acceleration of the 3-kg mass across the table?

   Work shown:

   $$f = 0$$
   $$T = m_1 a = m_2 (g - a)$$
   $$a = \left(\frac{m_2}{m_1 + m_2} \right) g = \frac{0.5}{3.5} g$$

Answer:

$a = 1.4 \text{ m/s}^2$

Requires clear, systematic approach and an ability to use algebra.

b. What is the tension in the string while the system is accelerating?

   Work shown: $T = m_1 a = 3 \times 1.4$

Answer:

$T = 4.2 \text{ N}$

c. If the mass starts from rest in the position shown in the diagram, what is the velocity of the 3-kg mass when it encounters the oil slick on the table?

   Work shown: $v_2 = v_0^2 + 2a\Delta x = 2 \times 1.4 \times 2$ m

Answer: $v = 2.37 \text{ m/s}$

Begin by using acceleration derived in part a, then use the time-independent kinematic equation over a distance of 2 m.

d. The oil provides friction between the block and the table, with coefficient $\mu_K = 0.3$. How far into the oil slick does it slide before coming to rest?

   Work shown:

$$f = mkm_1g$$
$$T = m_1(a + m_k g) = m_2(g - a)$$
$$a = \left(\frac{m_2 - m_k m_1}{m_1 + m_2}\right) g = \frac{0.5 - 0.3 \times 3}{3.5} g = -1.12 \text{ m/s}^2$$
$$d = (v^2 - v_0^2)/2g = -(2.37)^2/2(-1.12)$$

Answer: $d = 2.5$ m

This last part is rather hard. Not many students get it right. The solution requires repeating the calculations in a, b, and c, with additional horizontal force $f = m_k N$. $N = mg$ for 3-kg mass.

Hour Exam 1 Formula Sheet: Physics 101

Vector Components

$\Delta r = r_f - r_i$ Displacement $A_x = A \cos\theta$

$v = \frac{\Delta r}{\Delta t}$ Average Velocity $A_y = A \sin\theta$

$a = \frac{\Delta v}{\Delta t}$ Average Acceleration $A = \sqrt{A_x^2 + A_y^2}$

$\tan\theta = \frac{A_y}{A_x}$

$$v = v_0 + at$$
$$x = x_0 + v_0 t + \tfrac{1}{2} at^2 \quad \text{Kinematic equation}$$
$$v = \tfrac{1}{2}(v + v_0)$$
$$v^2 = v_0^2 + 2ax \qquad \text{Energy equation}$$

2D Kinematics

$a_x = 0$	$a_y = -g$
$v_x = v_{x0}$	$v_y = v_{y0} - gt$
$x = x_0 + v_{x0}t$	$y = y_0 + v_{y0}t - \tfrac{1}{2}gt^2$
	$v_y^2 = v_{y0}^2 - 2gy$

$$v_{ac} = v_{ab} + v_{bc} \qquad \text{Relative velocities}$$

Friction

$\Sigma F = ma$ Newton's II Law $f_s \le \mu_s N$ Static

$W = mg$ Weight $f_k = \mu_k N$ Kinetic

$g = 9.80 \text{m/s}^2$

Part 1—Multiple choice. *Circle the correct answer. If a calculation is required, you must show your work to receive credit. There is no partial credit in part 1.*

> With multiple-choice questions, it may be useful to work from the answers, especially if the solution to the problem is not obvious to you. Look for answers that seem approximately correct. Test them. Also look for answers that are reasonable, plausible. Eliminate the rest.

M1. (7 pts) A bicycle wheel of mass 4 kg and moment of inertia 6 kg/m² is spinning freely with an angular velocity of 8 rad/s. You gently grab hold of it with your hand and bring it to a stop. The shaft of the wheel does not move. How much work do you do on the wheel?

a. $W = 32$ J

b. $W = 48$ J

c. $W = 128$ J

d. $W = 192$ J

e. It depends on how you grab it.

Work shown:

$$KE_i = \frac{1}{2} I\omega^2$$
$$= 192 \text{ J}$$
$$KE_f = 0$$

Answer: $W = 192$ J

> This is a rotation work/energy problem. Begin by using the work-energy theorem. Find *KE* of rotating wheel, using the formula sheet. Because the final *KE* = 0, *KE* = work done. Once the physics is understood, this problem is simple number plugging. It is very often helpful to start plugging numbers into equations as soon as possible. Usually, it is easier to work with numbers than with variables.

M2. (7 pts) Two sacks of corn with masses $m_1 = 3$ kg and $m_2 = 6$ kg slide from rest down frictionless ramps until they collide. After the collision, they are both stuck together and not moving. If m_1 started from a height of $h_1 = 2$ m, what must have been the starting height of the other sack, h_2?

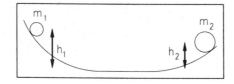

a. $h_2 = 0.5$ m
b. $h_2 = 1.0$ m
c. $h_2 = 1.4$ m
d. $h_2 = 2.0$ m

Work shown:

$$v_1 = \sqrt{2gh_1} \qquad v_2 = -\sqrt{2gh_2}$$

$$\sum P_{initial} = m_1 v_1 + m_2 v_2 = P_{final} = 0$$

$$m_1\sqrt{2gh_1} = m_2\sqrt{2gh_2} \qquad m_1\sqrt{2gh_1} = m_2\sqrt{2gh_2}$$

Answer: a

This is a case of inelastic collision demonstrating conservation of momentum. Begin by writing equations for velocities before collision, then apply conservation of momentum, with $P_{final} = 0$. This problem combines two different kinds of analysis, though each is simple. The problem requires good lateral reasoning to solve.

M3. (7 pts) A stiff spring is found to have compressed by 3 cm when it supports a mass of 7 kg. It is standing still in this compressed state. How much energy is stored in the compressed spring?

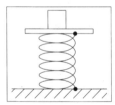

a. 0.031 J
b. 0.105 J
c. 1.03 J
d. 2.06 J

Work shown:

$$F = mg = k\Delta x$$
$$7 \times 9.8 = k \times 0.03$$
$$k = 2287 \text{ N}/\text{m}$$
$$KE_s = \frac{1}{2}k(\Delta x)^2 = \frac{1}{2}2287 \times (0.03)^2$$
$$= 1.029 \text{ J}$$

Answer: c

For this spring potential energy problem, use Hooke's law to find k. Use the formula sheet to find energy from PE_S. Solving this problem requires some imagination, because you are not given the spring constant, k, directly. Multiplying force by distance gives the wrong answer because the force is not constant.

M4. (7 pts) A communications satellite, launched into a circular orbit around the Earth, reaches an altitude of 400 km. What is the period of its orbit?

$(G = 6.67 \times 10^{-11} \text{ N m}^2/\text{kg}^2, \quad M_{\text{Earth}} = 6.0 \times 10^{24} \text{ kg}, \quad R_{\text{Earth}} = 6.4 \times 10^6 \text{ m})$

a. 79 s
b. 510 s
c. 5,100 s
d. 5,600 s
e. 86,000 s

Work shown:

$$a_c = V^2/R$$

$$F_g = ma_c = m\frac{v^2}{R} = G\frac{mM}{R^2}$$

$$V = \sqrt{\frac{GM}{R}}$$

$$\text{Period } T = \frac{2pR}{V} = 2p\ R\sqrt{\frac{R}{GM}}$$

$$= 2\pi\ \sqrt{\frac{R^3}{GM}} \qquad R = R_E + 4 \times 10^5 = 6{\cdot}8 \times 10^6$$

$$= 5569s$$

Answer: d

This gravitational force problem occurs commonly, so it is a good idea to memorize the result. First, find the gravitational force, then equate with mass multiplied by centripetal acceleration.

Part 2—Problems. *Show your work. Partial credit will be given if it is clear from your work that you are making progress toward the solution.*

P1. (18 pts) A grinding wheel with moment of inertia 0.4 kg/m² is designed to operate at 2500 r/min (revolutions per minute). Its motor provides a constant torque of 2 N m until it reaches the correct speed.

a. What is the angular acceleration of the wheel after it is turned on?
Work shown:

$$\alpha = \frac{\tau}{I} = \frac{2}{0.4}$$

Answer: $\alpha = 5$ rad/s²

b. Starting from rest, how long does it take to run up to its operating speed?
Work shown:

$$\omega_f = \omega_o + at$$

$$t = \frac{w_f}{a} = \frac{2500 \times 2p/60}{5} = \frac{262}{5}$$

Answer: $t = 52$ s

c. How many revolutions of the wheel are completed before it reaches operating speed?
Work shown:

$$\omega_f^2 = \omega_o^2 + 2\alpha\,\Delta\theta$$

$$\Delta\theta = \frac{\omega_f^2}{2\alpha} = 6854 \text{ rad}$$

$$= \frac{6854}{2\pi} \text{ turns}$$

Answer: $N = 1090$

> This rotational dynamics problem would be easy if it were linear motion instead of rotational. You should recognize the linear analogy. Do not miss this problem if you are short of time. In part a, begin by plugging numbers of torque and moment of inertia. In part b, convert ω into standard units and use the rotational kinematic equation. Finally, note that the answer for part c is time independent.

P2. (24 pts) A diving board is constructed from a uniform plank of wood of length $L = 10$ m and mass $M = 120$ kg. It is supported by a pivot at point B and a wire rope connected to the ground at its end A. A diver of mass $m = 80$ kg walks out along the board to make a dive.

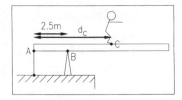

a. When the diver is standing directly over the pivot point B, what is the tension in the wire at A?

Work shown:

$$\sum \tau_B = 2.5 \times T - 2.5 \times Mg = 0$$

$$T = Mg = 120 \times 9.8$$

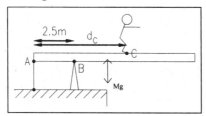

Answer: $T = 180$ N

This is a problem in rotational statics and dynamics. The mass of the diving board acts as its center. Begin by drawing a free-body diagram that shows all four forces. To solve part a, take the torques about B to simplify the resulting equation: $\sum \tau = 0$ for static $\equiv$ equilibrium *m*.

b. If the maximum tension withstood by the wire is 2500 N, how far along the board can the diver stand before the wire breaks?

Work shown:

$$\sum \tau_B = 2.5 \times T - 2.5 \times Mg - (d_c - 2.5)mg = 0$$

$$T = 2500 \text{ N at breakpoint}$$

$$d_c = 2.5 + \frac{2.5 \times 2500 - 2.5 \times Mg}{mg}$$

Answer: $d_c = 6.7$ m

Repeat the procedure used to solve part a, with the extra torque due to the diver. Get $T = 2500$ N and solve for d_c.

c. The diver now steps onto the board at a distance $d_c = 8$ m and the wire breaks, so that it no longer supports the board. The moment of inertia of the board about B is given by the formula $(7/48)ML^2$. What is the combined moment of inertia of the diver *and* board about B?

Work shown:

$$I_B = \frac{7}{48} ML^2 + (8 - 2.5)^2 m$$

$$1750 + 2420$$

Answer: $I_B = 4170$ kg/m^2

Use the definition $I = \sum m_i R_i^2$. Note that $I_{tot} = I_{board} + I_{diver}$.

d. What is the subsequent angular acceleration of the diver, standing at distance $d_c = 8$ m, *and* the board?

Work shown:

$$\sum \tau_B = -2.5 \times Mg - (8 - 2.5)mg$$

$$= -2940 - 4312 = -7252 = I_B \alpha$$

$$\alpha = \frac{-7252}{4170}$$

Answer: $\alpha = -1.74$ rad/s²

Set $T = 0$ and $\sum \tau_B$ as before, then use $\tau = I\alpha$ to find $\propto$. Even if you can't get part c, you can still do part d by assuming a number for I_B.

P3. (30 pts) A billiard ball of mass 0.2 kg and velocity $v_0 = 4$ m/s strikes a second ball of mass 0.2 kg, which is initially stationary as shown in the diagram. The balls move without friction in a horizontal x-y plane. You can ignore the effects of rolling. The first ball leaves the collision with a velocity of $v_1 = 3$ m/s at an angle of 35° as shown.

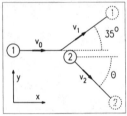

a. What is the *x* component of the *change* of the first ball's momentum during the collision?

Work shown:

$$\Delta p_x = m(v_1 \cos 35 - v_0)$$

$$\qquad\qquad 2.46 \quad 4$$

Answer: $\Delta p_x = -0.31$ kg m/s

Take components of v_1 in x and y. Δp_x = mass × change in x component of velocity.

b. What is the *y* component of the change of its momentum?

Work shown:

$$\qquad\qquad 1.72 \; m/2$$

$$\Delta p_y = m \qquad v_1 \sin 35$$

Answer: $\Delta p_y = 0.34$ kg m/s

Follow the same procedure as for the solution to part a.

c. If the collision lasts for 0.01 s, what is the magnitude of the average force acting between the balls during the collision?

Work shown:

$$\Delta p = \sqrt{\Delta p_x^2 + \Delta p_y^2} = \overline{F}\Delta t$$
$$0.46$$

Answer: $|\overline{F}| = 46$ N

The impulse-momentum theorem gives separate F_x and F_y components from parts a and b of the problem. Use Pythagoras to arrive at $|\overline{F}|$.

d. What is the angle θ at which the *second* ball leaves the collision?

Work shown:

$$\Delta P_1 + \Delta P_2 = 0 \text{ conservation of momentum}$$

$$\Delta p_{2x} = 0.31 \qquad \theta = \tan^{-1}\frac{-0.34}{0.31}$$

$$\Delta p_{2y} = -0.34$$

Answer: $\theta = 48°$

Because you are given neither the magnitude nor direction of v_2, you must get these from the conservation of momentum equation. Knowing

$$\Delta \xrightarrow{p_2} = m_2 \xrightarrow{v_2}, \text{ find } \theta \text{ using trig.}$$

e. What is the final velocity of the second ball?

Work shown:

$$\Delta p_2 = 0.46 \text{ kg m/s}$$
$$m_2 = 0.2 \text{ kg}$$
$$v_2 = \frac{0.46}{0.2} = 2.3 \text{ m/s}$$

Answer: $v_2 = 2.3$ m/s

Follow the same procedure as for part d, using Pythagoras to take the magnitude. Two-dimensional collision problems are complicated. You should not assume energy conservation unless it is explicitly stated.

Hour Exam #2 Formula Sheet: Physics 101

$$\Delta x = x_f - x_i \qquad v = \frac{\Delta x}{\Delta t} = \frac{x_f - x_i}{t_f - t_i} \qquad a = \frac{\Delta v}{\Delta t} = \frac{v_f - v_i}{t_f - t_i}$$

$$v = v_o + at \qquad x = v_{xo}t \qquad \Sigma F = ma$$

$$x = \tfrac{1}{2}(v + v_o)t \qquad v_y = v_{yo} - gt \qquad w = mg$$

$$x = v_o t + \tfrac{1}{2}at^2 \qquad y = v_{yo}t - \tfrac{1}{2}gt^2 \qquad f_s \le \mu_s N$$

$$v^2 = v_o^2 + 2ax \qquad v_y^2 = v_{yo}^2 - 2gy \qquad f_k = \mu_k N$$

$$W = (F\cos\theta)s \qquad W_{net} = KE_f - KE_i \qquad \text{kinetic energy}$$
$$KE = \tfrac{1}{2}mv^2$$

gravitational energy $\qquad$ conservation of energy $\qquad$ power
$$PE = mgh \qquad KE_i + PE_i = KE_f + PE_f \qquad P = \frac{W}{\Delta t}$$

spring energy
$$PE_s = \tfrac{1}{2}kx^2 \qquad W_{nc} = (KE_f - KE_i) + (PE_f - PE_i)$$

impulse $\qquad$ momentum
$$F\Delta t = \Delta p = mv_f - mv_i \qquad p = mv$$

$$m_1 v_{1i} + m_2 v_{2i} = m_1 v_{1f} + m_2 v_{2f}$$

$$\tfrac{1}{2}m_1 v_{1i}^2 + \tfrac{1}{2}m_2 v_{2i}^2 = \tfrac{1}{2}m_1 v_{1f}^2 + \tfrac{1}{2}m_2 v_{2f}^2$$

avg. angular velocity $\qquad$ avg. angular acceleration
$$\omega = \frac{\theta_2 - \theta_1}{t_2 - t_1} = \frac{\Delta\theta}{\Delta t} \qquad a = \frac{\omega_2 - \omega_1}{t_2 - t_1} = \frac{\Delta\omega}{\Delta t}$$

$$\omega = \omega_o + \alpha t \qquad\qquad\qquad a_c = \frac{v^2}{r} = r\omega^2$$

rotational kinematic equ. $\qquad$ tangential velocity
$$\theta = \omega_o t + \tfrac{1}{2}\alpha t^2 \qquad v_t = r\omega \qquad F_c = m\frac{v^2}{r}$$

rotational energy equ. $\qquad$ tangential acceleration $\qquad$ Universal law of Gravitation
$$\omega^2 = \omega_o^2 + 2\alpha\theta \qquad a_t = r\alpha \qquad F = G\frac{m_1 m_2}{r^2}$$

$$\theta = \tfrac{1}{2}(\omega_o + \omega)t$$

rotational Newton's II $\qquad$ moment of inertia $\qquad$ Torque
$$\Sigma\tau = I\alpha \qquad I \equiv \Sigma mr^2 \qquad \tau = Fd$$

THIRD HOUR EXAM

Part 1—Multiple choice. *Circle the correct answer. If a calculation is required, you* must *show your work to receive credit. There is no partial credit in part 1.*

M1. (7 pts) A metal pot contains water that is boiling on a red-hot barbeque grill. What mechanisms of heat transfer are active between the hot coals and the steam produced?

a. radiation and convection
b. convection and conduction
c. conduction and radiation
d. convection, conduction, and radiation

Work shown:

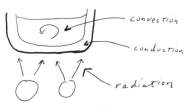

Answer: d

The pot is metal, so it needs conduction for heat to cross it. The water is liquid, which usually carries heat by convection. The red-hot coals cannot touch the pot (which is on a grill), so the system must use radiation.

M2. (7 pts) A fisherman's boat tows a fishing line carrying four spherical polystyrene floats, each of radius 0.02 m, which remain submerged. The boat is traveling with a constant velocity of 15 m/s. He detects a tension of 0.26 N in his line due to the floats (and has not yet caught a fish). What is the viscosity of the seawater?

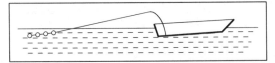

a. 0.0029 Pa • s
b. 0.0046 Pa • s
c. 0.011 Pa • s
d. 0.046 Pa • s

Work shown:

$$F_D = 6\pi\eta Rv \times 4 \text{ floats}$$

$$\eta = \frac{F_D}{24\pi Rv} = \frac{0.26 \text{ N}}{24\pi\, 0.02 \times 15} = 0.0115 \text{ Pa}\cdot\text{s}$$

Answer: c

This is a viscosity problem. The viscous drag on a moving sphere is given by Stoke's law. Because there are four spheres, the drag force = 1/4 of the tension in the line.

M3. (7 pts) 2500 J of heat are absorbed by a mechanical heat engine while it lifts a mass of 20 kg through 2 m against gravity. What is the change of internal energy of the engine?

a. 2900 J
b. 2100 J
c. 0 J
d. −2100 J
e. −2900 J

Work shown:
$\Delta U = Q - W$ (1st law)
Q = heat in = +2500 J
W = work done = $20 \times 9.8 \times 2 = 392$ J

Answer: b

The applicable principle here is the first law of thermodynamics. Find the work output by the machine, then subtract it from the heat input.

M4. (7 pts) A thermometer is made from glass (which has no thermal expansion) and 15 cm³ of a red liquid with a *volume* expansion coefficient of 4×10^{-4} K⁻¹. The tube has a *radius* of 1 mm. The distance the liquid rises in the tube is read on a calibrated scale. How far apart should the "0°C" and "100°C" marks be placed on its scale?

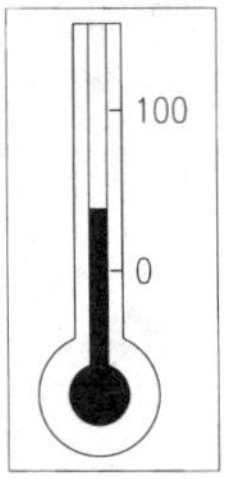

a. 19.1 cm
b. 4.78 cm
c. 1.27 cm
d. 0.19 cm

Work shown:

$$\Delta V = \beta V_0 \Delta T$$
$$= 4 \times 10^{-4\circ}C^{-1} \times 15 \times 100$$
$$= 0.6 \text{ cm}^3$$
$$L = \frac{\text{volume}}{\text{area}} = \frac{0.6}{\pi (0.1)^2} = 19.1 \text{ cm}$$

Answer: a

This is a volume expansion problem. Begin by finding the increase in volume of the liquid for a 100°C rise, then find the area of the tube, and divide volume by area.

Part 2—Problems. *Show your work. Partial credit will be given if it is clear from your work that you are making progress toward the solution.*

P1. (24 pts) A balloon of volume 0.2 m³ contains helium gas (0.004 kg/mol) at atmospheric pressure at a temperature of 320 K. The fabric of the balloon itself has negligible mass, but it is attached to a long string, which has a mass-per-unit length of 0.005 kg/m.

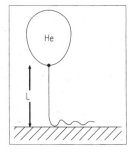

a. What mass of helium does the balloon contain?

Work shown:

$$PV = nRT$$
$$n = \frac{PV}{RT} = \frac{1.01 \times 10^5 \times 0.2}{8.31 \times 320} = 7.60 \text{ mol}$$
$$\text{mass} = 7.60 \times 0.004$$

Answer: $m = 0.0304$ kg

Use the gas law to get the number of moles, then multiply by mass per mole.

b. A child holds the string so that there is only a short length of string between her hand and the balloon. What is the tension in the string, measured at her hand? The density of the air is 1.2 kg/m³.

Work shown:

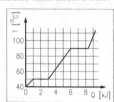

$$F_B = \rho g V = 1.2 \times 9.8 \times 0.2 = 2.35 \text{ N}$$
$$T = F_B - mg = 2.35 - 0.0304 \times 9.8$$

Answer: $T = 2.05$ N

> Knowing the volume of the balloon, calculate buoyant force. Draw a free-body diagram of the forces, and find the connection between *B*, *T*, and *mg*.

c. The child now releases the balloon, which rises until the weight of the length of string above the ground balances the upward force, as in the picture. How high is the balloon?

Work shown:

$F_B = 2.35$ N still

$mg = F_B$ including mass of string

$$m = 0.24 \text{ kg}$$

$$\frac{0.304 \text{ kg He}}{0.209\text{kg string}}$$

$$L = \frac{0.209 \text{ kg}}{0.005 \text{ kg} / \text{m}}$$

Answer: $L = 41.9$ m

> This problem requires some intuition, because quantities like "mass per unit length" will not have been introduced explicitly. Nevertheless, the problem is easy. Knowing *T* from part b, find the mass of the string. Length = mass divided by mass per unit length.

P2. (24 pts) 0.025 kg of a pure organic solid is heated while its temperature, T, is monitored as a function of the total heat added, Q, as shown in the picture. During the course of the experiment it melts and then boils.

a. At what temperature does the compound boil?

Answer: $T = 90°C$

b. What is its latent heat of fusion?

   Work shown:
$$\frac{2 \text{ kJ}}{0.025 \text{ kg}}$$

Answer: $L_f = 80$ kJ/kg

> This is a latent heat problem. You must be able to recognize changes of state from a graphical signature. In part a, the changes of state are the "flats" on the curve. $T_{boil} > T_{melt}$ always. In part b, fusion is melting = lower "flat." Read off the heat input from the graph and divide by mass for latent heat per kilogram.

c. A test tube containing 0.025 kg of the boiling liquid is placed in 0.05 kg of water at 30°C. It is observed that the liquid *partly* solidifies. What is the final temperature of the water? Neglect any heat losses due to the containers. The specific heat of water is 4186 J/kg • K.

   Work shown: Partly solidifies $\Rightarrow$ solid + liquid. Must be T_m.

Answer: $T_{water} = 50°C$

> This is a trick question. Note the word *partly* in the question. A partly solidified material must still be at its melting point.

d. What mass of solid compound has formed after equilibration with the water?

   Work shown: $Q_{water} = 0.05 \text{ kg} \times 4186 \text{ J / kg K} \times (50 - 30) = 4186 \text{ J}$

   $Q_{organic} = 3 \text{ kJ (specific heat)} + \Delta Q_{latent}$

   $\Delta Q_{latent} = 1186 \text{ J} = mL$

   $m = \dfrac{1186}{L} = 0.0148 \text{ kg of solid}$

Answer: $m_{solid} = 0.0148$ kg

> This question was designed for students who had observed these effects in lab. To solve part d, find the heat gained by water = heat lost by boiling liquid. We know the final temperature from part c. Next, heat lost = specific heat part + latent heat part. The latent part gives the mass of the solid.

P3. (24 pts) Your bathtub is filled with water to a depth of 0.3 m. To empty the tub, you pull on a light string attached to a small, massless plug, which initially covers the drain. The drain is round, with an inner diameter of 0.04 m, and drops down into an open sewer (at atmospheric pressure) 0.8 m below. The water is clean and can be considered to be an ideal nonviscous incompressible fluid of density 1000 kg/m³.

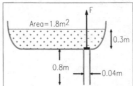

a. How much force, F, must you exert on the string to remove the plug?

Work shown:

$$P = \rho g h = \frac{F}{A} = \frac{1000 \times 9.8 \times 0.3}{0.00126}$$

$$A = \pi r^2 = \pi (0.02)^2 = 0.00126$$

Answer: $F = 3.69$ N

A problem in fluid statics and dynamics. Note that fluids are ideal when they have laminar flow explained by Bernoulli's equation. First, find the pressure at a depth of 0.3 m. $P = $ force/area.

b. Immediately after you remove the plug, what is the *initial* velocity of the water traveling down the drain?

Work shown:

$$\rho g h = \frac{1}{2} \rho V^2$$

$$V = \sqrt{2gh}$$

Answer: $V_{d1} = 2.42$ m/s

Bernoulli's equation relates velocity to depth.

c. Immediately after you remove the plug, at what rate does the water level in the tub recede?

Work shown:

$$V_{tub} = \frac{A_{drain}}{A_{tub}} \times V_{d1} = \frac{0.00126}{1.8} \times 2.42$$

Answer: $V_{tub} = 0.00170$ m/s

Knowing velocity, we know the volume flow rate. The volume flow rate = $V_{tub} \times$ top surface area.

d. Moments later, the drain pipe has become filled with water. What is the velocity of the water down the drain now? You may assume that the water level in the tub has not yet dropped significantly.

Work shown:

$$\rho g h = \frac{1}{2} \rho v^2$$

$$V = \sqrt{2gh}$$

$$h = 1.1 \text{ m}$$

Answer: $V_{d2} = 4.64$ m/s

Fluid is nonviscous, so there are no drag forces (Poiseuille). Conservation of energy still applies per Bernoulli equation. Bernoulli equation with new depth = 1.1 m.

Hour Exam #3 Formula Sheet: Physics 101

density
$$\rho \equiv \frac{M}{V}$$

pressure
$$P = \frac{F}{A}$$

$$P = P_a + \rho\, gh$$

buoyant force
$$B = \rho_f\, gV$$

surface tension
$$\gamma \equiv \frac{F}{L}$$

$$h = \frac{2\gamma}{\rho gr} \cos\phi$$

continuity
$$Q = \frac{\Delta V}{\Delta t} = A_1 v_1 = A_2 v_2$$

Bernoulli
$$P + \tfrac{1}{2}\rho\, v^2 + \rho\, gy = \text{const}$$

Poiseuille
$$Q = \frac{\Delta V}{\Delta t} = \frac{(P_1 - P_2)\pi R^4}{8L\eta}$$

Stokes
$$F_{drag} = 6\,\pi\eta\, rv$$

$$v = \sqrt{2gh}$$

$$\Delta L = \alpha L_o \Delta T$$

$$\Delta A = \gamma A_o \Delta T$$

$$\Delta V = \beta V_o \Delta T$$

$$PV = nRT$$

$$P = \frac{2}{3}\left(\frac{N}{V}\right)\left(\tfrac{1}{2}m_o \overline{v^2}\right)$$

$$\tfrac{1}{2}m_o \overline{v^2} = \tfrac{3}{2}kT$$

$$v_{rms} = \sqrt{\frac{3kT}{m_o}} = \sqrt{\frac{3RT}{M}}$$

$$T_c = T - 273.15$$

$$T_F = \tfrac{9}{5}T_c + 32$$

$$Q = mc\,\Delta T$$

$$H = \frac{Q}{\Delta t} = kA\left(\frac{T_2 - T_1}{L}\right)$$

$$\frac{Q}{\Delta t} = hA\Delta T$$

$$Q = mL$$

Stefan's Law
$$P = \sigma A e T^4$$

$$P_{net} = \sigma A e (T^4 - T_o^4)$$

$$W = F\,\Delta x = P\,\Delta V$$

internal energy
$$\Delta U = U_f - U_i = Q - W$$

$$U = \tfrac{3}{2}NkT = \tfrac{3}{2}nRT$$

$$\Delta S = \frac{Q}{T}$$

$$\epsilon = \frac{W}{Q} < 1 - \frac{T_C}{T_H}$$

$g = 9.8$ m/s^2

$R = 8.31$ J/(mol K) $= 0.0821$ (L atm)/(mol K)

$k = 1.38 \times 10^{-23}$ J/K

$N_A = 6.023 \times 10^{23}$ molecules/mol

$\sigma = 5.67 \times 10^{-8}$ W/(m^2 K^4)

$\rho_{water} = 1000$ kg/m^3

1 cal = 4.186J

1 atm = 1.01×10^5 Pa

1 liter = 10^{-3} m^3

1 liter atm = 101J

1 inch = 25.4mm

CHAPTER 25

UNIVERSITY OF OKLAHOMA

PHYSICS 2414: INTRODUCTORY PHYSICS (NON-CALCULUS)

Michael Strauss, Professor

THREE MIDTERM EXAMS ARE GIVEN IN THIS COURSE, AS WELL AS A CUMULATIVE FINAL. An example of one midterm and the final are included here.

MIDTERM

Each multiple-choice question is worth 5 points.

1. A man pushes a stalled car to get it started. During the time when the man is speeding up, which of the following statements is true?
   a. Neither the car nor the man exert any force on the other.
   b. The man exerts a force on the car, but the car does not exert any force on the man.
   c. The magnitude of the force that the man exerts on the car is equal to the magnitude of the force the car exerts on the man.
   d. The magnitude of the force that the man exerts on the car is less than the magnitude of the force the car exerts on the man, but neither force is zero.
   e. The magnitude of the force that the man exerts on the car is greater than the magnitude of the force the car exerts on the man, or the car would not accelerate forward, but neither force is zero.

   Answer: c

   The student should know that Newton's third law applies between two objects even when the objects are accelerating.

2. A rock is suspended from a string and moves downward at a *constant speed.* Which statement is true concerning the tension in the string if air resistance is ignored?

a. It is zero.

b. It is equal to the weight of the rock.

c. It is greater than the weight of the rock.

d. It points downward.

e. It is less than the weight of the rock.

Answer: b

At constant speed, Newton's second law says that the sum of the forces is zero. The only forces on the rock are its weight and the tension, so the tension should be the same as the weight.

3. Before CDs were invented, we used to listen to music on thin pieces of plastic with groves in them called records. Suppose you put a penny on a record and let it spin clockwise on the turntable. After turning the turntable off, it is still spinning, but is slowing down. Which arrow shows the direction of the acceleration of the penny at point *P* during this time when the turntable is spinning but slowing down?

Answer: d

To solve this, the student must understand three things: (1) the centripetal acceleration is toward the center, (2) the tangential acceleration is in the opposite direction of motion, (3) when vectors from (1) and (2) are correctly added, the answer will be as given.

4. Which is an accurate statement regarding satellites in orbit?

a. The satellite has no forces acting on it, and so it is weightless.

b. The velocity required to keep a satellite in a given orbit depends on the mass of the satellite.

c. It is possible to have a satellite traveling at either a high or low speed in a given circular orbit.

d. The period of revolution of a satellite moving about the earth does not depend on the size of the orbit it travels.

e. A satellite in a large-diameter circular orbit will have a longer period of revolution about the earth than will a satellite in a smaller circular orbit.

Answer: e

This answer can be determined by realizing that the other four options are incorrect based on class discussion and equations we have derived and discussed regarding circular orbits. Or, it can be worked out in detail using the equations that follow.

$$Gm_, m_E / r^2 = m_, v^2 / r$$
$$v = \sqrt{Gm_E / r}$$

$$T = 2\pi r / v = 2\pi \sqrt{r / Gm_E}$$
so the period is larger for a larger r.

5. Two satellites A and B of the same mass are going around the earth in concentric circular orbits. The distance of satellite B from earth's center is twice that of satellite A. What is the ratio of the centripetal force acting on B to that acting on A?

   a. 1/8
   b. 1/2
   c. 1
   d. 1/4
   e. $\sqrt{\frac{1}{2}}$

Answer: d

This problem tests that the student knows the centripetal force on a satellite is given by $Gm_1 m_E / r_2$, and whether the student understands what happens to the force when the radius doubles, and whether the student can take ratios. This is the type of problem I give to see if the student can really think through the concepts we have discussed.

$$F_A = Gmm_E / r^2$$
$$F_B = Gmm_E / (2r)^2$$
$$F_B / F_A = 1/4$$

6. (22 pts) You are standing on a scale in a stationary elevator, which shows your weight to be 710 N. (That is, the force of your body pushing down on the scale is 710 N.) As the elevator accelerates upward, you notice that the scale shows you have an apparent weight of 860 N. (Now, you are pushing down on the scale with a force of 860 N.)

   a. What is your mass?
   b. When the elevator is accelerating upward, what is the magnitude of the normal force of the scale pushing up on you?
   c. What is the magnitude of the acceleration?
   d. If the elevator is moving at a constant velocity upward, what does the scale read?

Answers:

a. $mg = 710$ N

   $m = 710$ N$/9.80$ m/s$^2 = 72$ kg

b. By Newton's third law, 860 N

c.

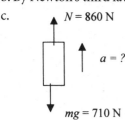

$\Sigma F = ma$

$N - mg = ma$

$a = (860$ N $- 710$ N$)/72$ kg $= 2.1$ m/s^2

d.

$\Sigma F = ma$

$N - mg = ma$

$n = mg = 710$ N

Always draw force diagrams when using Newton's second law. The apparent weight is usually given by the normal force. When an object has a constant velocity, the net force on that object is zero.

---

7. (33 pts) A block on an inclined plane is connected to a second block by a massless string hanging over a massless, frictionless pulley, as shown in the figure at the right. The mass m_1 on the incline experiences a frictional force. It is observed that when the block m_2 is released from rest it falls 1.40 m in 1.20 s.

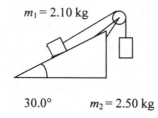

$m_1 = 2.10$ kg

30.0° $m_2 = 2.50$ kg

   a. What is the acceleration of m_2?

   b. What is the tension in the string?

   c. What is the coefficient of kinetic friction between m_1 and the incline?

Answers:

a. $x - x_0 = v_0 t + (1/2)at^2$

   $a = 2(x - x_0)/t = 2(1.40$ m$)/(1.20$ s$)^2 = 1.94$ m/s^2

b.

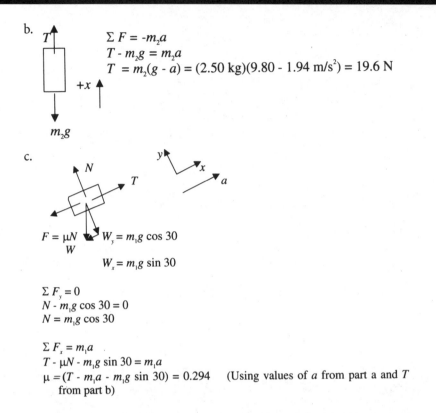

$\Sigma F = -m_2 a$

$T - m_2 g = m_2 a$

$T = m_2(g - a) = (2.50 \text{ kg})(9.80 - 1.94 \text{ m/s}^2) = 19.6 \text{ N}$

c.

$F = \mu N$ $W_y = m_1 g \cos 30$

$W_x = m_1 g \sin 30$

$\Sigma F_y = 0$

$N - m_1 g \cos 30 = 0$

$N = m_1 g \cos 30$

$\Sigma F_x = m_1 a$

$T - \mu N - m_1 g \sin 30 = m_1 a$

$\mu = (T - m_1 a - m_1 g \sin 30) = 0.294$ (Using values of a from part a and T
   from part b)

Always draw force diagrams when solving problems with Newton's second law. When dealing with more than one object, draw force diagrams for each object. Always separate vectors into components and write one equation for the forces and accelerations along x and one along y. It is easier to solve the problem if you choose your axes so that the acceleration in one direction is zero. That is why I choose the x and y axes tilted in this problem.

8.(25 pts) You take a home-made "accelerometer" to Six Flags over Texas during the OU-Texas weekend. This accelerometer consists of a metal nut attached to a string and connected to a protractor, as shown in the figure. While riding a roller coaster that is moving at a uniform speed around a circular path, you hold up the accelerometer and notice that the string is making an angle of 55° with respect to the vertical, as shown.

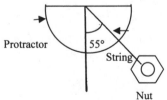

a. What is the centripetal acceleration of the roller coaster?

b. How many g's are you experiencing in the radial direction (1 g = 9.80 m/s²)?

c. If the roller coaster track is turning in a radius of 80.0 m, how fast are you moving?

Answers:

a.

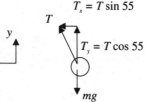

$T_x = T \sin 55$

$T_y = T \cos 55$

mg

$\Sigma F_y = 0$

$T \cos 55 - mg = 0$

$T = mg/\cos 55$

$\Sigma F_x = ma_c$

$T \sin 55 = ma_c$

$a_c = g \sin 55/\cos 55 = g \tan 55 = (9.80 \text{ m/s}^2) = 14 \text{ m/s}^2$

b. $(14 \text{ m/s}^2)(1 \text{ g}/9.80 \text{ m/s}^2) = 1.4$ g's

c. $a_c = v^2/r$

$v = \sqrt{a_c r} = \sqrt{(14 \text{ m/s}^2)(80.0 \text{ m})} = 33 \text{ m/s}^2$

> Again, always draw a force diagram, use vector components, and write equations in the *x* and *y* direction separately.

FINAL EXAM

Each multiple-choice question is worth 5 points.

1. You get so much new stuff during this holiday season that you are forced to rent a trailer to bring it all to OU. You get in your car and think about the forces acting on the car and the trailer as you begin to accelerate. Why do the car and the trailer accelerate forward?

   a. Because the force of the car pulling forward on the trailer is slightly greater than the force of the trailer pulling backward on the car.

   b. Because the car gives the trailer a quick tug during which the force of the car on the trailer is momentarily greater than the force of the trailer pulling backward on the car.

   c. Because the car weighs more than the trailer so it has more momentum.

   d. Because the force of the car pulling forward on the trailer is equal to the force of the trailer pulling backward on the car, but the frictional force on the car is forward and large, whereas the frictional force on the trailer is backward and small.

e. Because the force of the car pulling forward on the trailer is equal to the force of the trailer pulling backward on the car, but the motor in the car is powerful enough to pull both the car and the trailer.

Answer: d

> The student should know that the car moves because of the frictional force of the tires on the road, and that Newton's third law is always applicable between two objects even during acceleration.

2. You drop a heavy steel ball from a certain height. At the same time, you throw a light plastic ball horizontally from the same height. Neglecting air resistance, which ball will hit the ground first, and which one will have the greater velocity when they hit?

Hit First	Greater Velocity
a. Hit at the same time	Steel
b. Hit at the same time	Plastic
c. Hit at the same time	They have the same velocity
d. Steel	Plastic
e. Steel	Steel

Answer: b

> I am testing two concepts. Does the student know that the acceleration of gravity in the vertical direction is constant and the same for all objects? If so, then the two objects hit the ground at the same time. The second concept is whether the student knows that the total velocity is the vector sum of the vertical and horizontal components of velocity, so if the two objects have the same vertical velocity, and the second has a greater horizontal velocity, the second must also have the greater total velocity.

3. You put a piece of iron on top of a piece of Styrofoam and the two float in a glass container partially filled with water. You then take the iron from the top of the Styrofoam and instead suspend it by a thread below the Styrofoam, so that the iron is in the water, but the iron and Styrofoam together still float. What happens to the water level in the glass container compared with when the iron was sitting on top of the Styrofoam?

a. It drops.

b. It stays the same.

c. It rises.

d. More information is needed.

Answer: b

The only two forces on the floating Styrofoam and iron are its weight and the buoyant force of the water. Because there is no acceleration, the magnitude of these forces is the same whether the iron is on top or suspended. Because the buoyant force of water is equal to the weight of the water displaced and the buoyant force is the same in both cases, the water level will be the same in both cases.

4. You put a penny on a circular turntable, and watch it rotate at a constant speed as illustrated in the following diagram. While it is rotating, you think about all of the various vectors associated with this motion. Which of the sets of vectors below best describes the velocity, acceleration, and net force acting on the cylinder at the point indicated in the diagram?

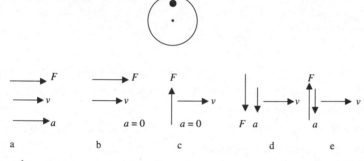

Answer: d

Testing whether the student understands circular motion, centripetal force, and acceleration.

5. You are designing a soapbox derby race car that will race down a hill using only gravity to produce acceleration. Suppose the rules stipulate a maximum weight for the car including the wheels, but do not stipulate how to distribute the weight. Ignore friction and air resistance. How do you design the wheels on your car to get down the hill quickest? (For this problem, all "heavy" wheels have the same mass, and all "light" wheels have the same mass.)

a. Make the wheels heavy thin disks.

b. Make the wheels light thin disks.

c. Make the wheels heavy hoops (like bicycle wheels).

d. Make the wheels light hoops (like bicycle wheels).

e. It doesn't matter how you design the wheels.

Answer: b

To answer this correctly, the student must realize that mechanical energy is conserved. The more mechanical energy that goes into the kinetic energy of the spinning wheels, the less available for translational kinetic energy (forward motion). Because the light thin wheels have a smaller moment of interia, they will have less rotational kinetic energy for the same translational speed.

6. When tuning your guitar, you find that you must increase the tension in one of the strings by a factor of four. What happens to the fundamental frequency of the string when you do this?

   a. It decreases by a factor of 4.

   b. It increases by a factor of 4.

   c. It stays the same.

   d. It decreases by a factor of 2.

   e. It increases by a factor of 2.

Answer: e

> The fundamental frequency is proportional to the wave velocity on the string, which is proportional to the square root of the tension, so when the tension increases by a factor of four, the frequency increases by a factor of two.

7. You throw a golf ball to the right so that it collides with a bowling ball that is initially at rest. The golf ball bounces back elastically so that after the collision it has almost the same velocity to the left that it initially had to the right. Which object has the greater change in momentum during the collision, and which object has the largest momentum after the collision?

Greatest change in momentum	Largest momentum after collision
a. Golf ball	Golf ball
b. Golf ball	Same for both balls
c. Bowling ball	Same for both balls
d. Same for both balls	Golf ball
e. Same for both balls	Bowling ball

Answer: e

> In any collision between two objects, the change in momentum of each of the objects is the same, because $F = dp/dt$, and the forces between the objects are the same by Newton's third law. The golf ball had to change its momentum from $+mv$ to $-mv$, or a difference of $2mv$, so the bowling ball acquired $2mv$ of momentum, because it started at rest. Consequently, the bowling ball has almost twice the momentum of the golf ball.

8. While doing surgery, you notice that one of your patient's arteries is partially blocked by deposits along the artery wall, so that the diameter of the artery is smaller where the deposits are. How does the velocity of the blood and the fluid pressure of the blood in the part of the artery that is blocked compare with the rest of the artery?

Velocity	Fluid Pressure
a. Slower than the rest of the artery	Greater than the rest of the artery
b. Slower than the rest of the artery	Lower than the rest of the artery
c. Faster than the rest of the artery	Greater than the rest of the artery
d. Faster than the rest of the artery	Same as the rest of the artery
e. Faster than the rest of the artery	Lower than the rest of the artery

Answer: e

Velocity for a flowing fluid increases with a smaller cross-sectional area, and fluid pressure decreases with increased velocity.

9. Resonance occurs in harmonic motion when
   a. the system is overdamped.
   b. the system is critically damped.
   c. the energy in the system is a minimum.
   d. the driving frequency is the same as the natural frequency of the system.
   e. the energy in the system is proportional to the square of the motion's amplitude.

Answer: d

This is probably the easiest problem on the test. I am just checking if the student knows the definition of resonance and various terms like "damped," and so forth.

10. Which of the following is *not* true?
    a. Light travels much faster than sound.
    b. Sound waves can travel through a vacuum.
    c. Sound waves are longitudinal pressure waves.
    d. The pitch of a sound is closely related to the frequency of the wave.
    e. When a vibrating string produces a sound wave in air, the frequency of the string vibration is the same as the frequency of the sound wave traveling in air.

Answer: b

Testing if the student knows some of the properties of light and sound waves. Simply testing for factual information.

11. (30 pts) You get some wood or some Hot Wheels track and build a ramp as shown in the figure. The first part of the ramp is shaped like a circle with radius 0.75 m (point *A* is at the bottom of the circle) and the second part ends in a horizontal section at point *B*. The track is on a table 1.5 m above the floor. You spray silicon on

the track so it is basically frictionless, and you release a 0.45-kg block from the top of the ramp 0.75 m above the surface of the table.

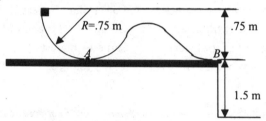

a. What is the speed of the block at point A?

b. Which direction is the acceleration when the block is at point A?

c. What is the normal force the ramp exerts on the block at point A?

d. What is the speed of the block at point B?

e. After the block leaves the ramp at point B, how long does it take for it to hit the floor?

f. How far from the end of the ramp does the block land on the floor?

Answers:

a. There are no nonconservative forces, so mechanical energy is conserved.

$$mgh = (1/2)mv^2$$
$$v = \sqrt{2gh} = \sqrt{2(9.8\text{m}/\text{s}^2)(0.75\text{m})} = 3.8\text{m}/\text{s}$$

b. Upward.

c. N $\Sigma F = ma = ma_c = mv^2/r$
$N - mg = mv^2/r$
$N = m(v^2/r + g) = (0.45\text{ kg})((3.8\text{ m/s}^2)/(0.75\text{ m}) + 9.80\text{ m/s}^2) = 13\text{ N}$

d. Same as at point $A = 3.8$ m/s.

e. $y - y_0 = v_{0y} + (1/2)gt^2$

$t = \sqrt{2(y - y_0)/g} = \sqrt{2(1.5\text{ m})/(9.80\text{ m}/\text{s}^2} = 0.55$ s

f. $x - x_0 = v_x t = (3.8\text{ m/s})(0.55\text{ s}) = 2.1$ m.

Always draw a force diagram when solving problems with Newton's second law. Remember that centripetal acceleration is always toward the center of the circle.

12. (30 pts) Unfortunately, you are in a car accident. You are convinced that the other person was speeding and the accident is not your fault. You were going north at 25 mi/h, and the other person was going east where the speed limit was 25 mi/h. After the collision, your cars stuck together and traveled 12 m exactly at an angle of 40° north of east. You find out that the coefficient of friction of all the tires on the road

is 0.40. You also know that your car has a mass of 1200 kg and the other person's car has a mass of 900 kg. (1 mi = 1.61 km.)

a. What was the frictional force of the tires on the road?
b. How much work was done by friction to stop the cars?
c. What was the velocity of the cars immediately after the collision?
d. What was the other person's velocity before the collision?
e. Was the other person speeding?

Answers:

a. (25 mi/h)(1.61 km/mi)(1000 m/km)(1 h/3600 s) = 11.1 m/s.

Let M = total mass = 2100 kg, m = mass of other car = 900 kg.

$$F = \mu N = \mu M g = (0.40)(2100)(9.8 \text{ m/s}^2) = 8200 \text{ N}$$

b. $W = Fd \cos \theta = (8200 \text{ N})(12 \text{ m})(\cos 180) = -9.9 \times 10^4 \text{ J}$

c. Friction is a nonconservative force, so we look at the change in mechanical energy.

$$W_{NC} = \Delta U + \Delta K = 0 + (1/2) M v_f^2 - (1/2) M v_i^2 = 0 - (1/2) M v_i^2$$

$$v = \sqrt{-2W_{NC} / M} = \sqrt{2(9.9 \times 10^4 \text{ J} / 2100 \text{ kg}} = 9.7 \text{ m/s}$$

d. Look at conservation of momentum in the eastern direction only (v_E):

$v_E = v \cos 40 = (9.7 \text{ m/s})(\cos 40) = 7.4 \text{ m/s}$

Use conservation of momentum in the eastern direction:

$m v_i = M v_E$

$v_i = M v_E / m = (2100 \text{ kg})(7.4 \text{ m/s})/(900 \text{ kg}) = 17 \text{ m/s}$

e. Yes, because 17 m/s is greater than 11 m/s (the speed limit).

> Momentum is a vector and so is conserved independently in two dimensions. In this problem, we only have to consider one direction. Remember to always use proper units (change miles per hour to meters per second to compare in this problem).

13. (25 pts) You are hanging Christmas ornaments outside of your house. Your big, gaudy Santa Claus is hanging from a beam and wires as shown. The beam has a length of 1.25 m and a mass of 5.00 kg, and Santa has a mass of 2.50 kg.

a. What is the tension in the top wire?

b. What are the vertical and horizontal components of the force of the house on the beam?

Answers:

a. $\sum \tau = mgl + Mgl/2 - Tl\sin 30 = 0$

$T = (mgl + Mgl/2)/(l\sin 30)$

$= \{(2.5 \text{ kg})(1.25 \text{ m}) + (5.00 \text{ kg})(1.25/2.0 \text{ m})\} (9.80 \text{ m/s}^2)/\{(1/25 \text{ m})(\sin 30)\} = 9.80 \text{ N}$

Look at the torques around the beam/house junction (l = 1.25 m).

b. $\sum F_x = 0$

$T\sin 30 - F_x = 0$

$F_x = T\sin 30 = (9.80 \text{ N})\sin 30 = 4.90 \text{ N to the left}$

$\sum F_y = 0$

$-mg - Mg - T\cos 30 + F_y = 0$

$F_y = mg + Mg + T\cos 30$

$= (2.50 \text{ kg})(9.80 \text{ m/s}^2) + (5.00 \text{ kg})(9.80 \text{ m/s}^2) - (9.80 \text{ N})\cos 30$

$= 65.0 \text{ N upward}$

Always draw a force diagram when using Newton's second law. You can choose the most convenient axis available when considering torques.

14. (25 pts) You decide to amaze your friends with stupid physics tricks. So, at a party you get a keg of root beer (because there may be underage people present) and poke a hole in the top. You then take a 13-m-long hose and put it in the hole, and fill up the hose with root beer. The radius of the top of the keg is 21 cm and the radius of the hose is 0.20 cm. Assume root beer has the same density as water (1000 kg/m³).

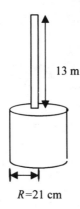

13 m

R=21 cm

a. What is the mass of the root beer in the hose?

b. What is the gauge pressure of the root beer at the bottom of the hose?

c. What is the force from the root beer on the lid of the barrel?

d. This force is equivalent to how many 75-kg (165-lb) people standing on the lid? Do you think the barrel will burst?

Answers:

a. $\rho = m/V = m/\pi r^2 h$

  $m = \rho \pi r^2 h = (1000 \text{ kg/m}^3) \, \pi \, (0.002 \text{ m})^2 \, (13 \text{ m}) = 0.16 \text{ kg}$

b. $P = \rho g h = (1000 \text{ kg/m}^3) \, (9.80 \text{ m/s}^2)(13 \text{ m}) = 1.3 \times 10^5 \text{ Pa}$

c. $P = F/A$

  $F = PA = (1.3 \times 10^5 \text{ Pa}) \, \pi \, (.21 \text{ m})^2 = 1.8 \times 10^4 \text{ N}$

d. $(1.8 \times 10^4 \text{ N})/[(75 \text{ kg})(9.80 \text{ m/s}^2)] = 24$ people

  Yes, it would most likely burst!

15. (30 pts) You have a spring at home and want to determine the spring constant of the spring. You hang the spring from the ceiling and it is 20.0 cm long. Then you hang a 1.10-kg weight on the end of the string and it stretches to 31.0 cm.

a. What is the spring constant of the spring?

b. You recall from physics class that a spring oscillating in a vertical direction behaves exactly like a spring oscillating in a horizontal direction, so you pull the mass 5.00 cm from the equilibrium position to watch it oscillate. What will be the maximum velocity of the oscillating mass?

c. When the displacement is 2.50 cm from the equilibrium position, what is the velocity of the mass?

d. How long will it take for the spring to oscillate five times?

Answers:

a.

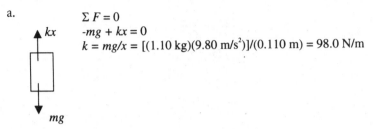

$\Sigma F = 0$

$-mg + kx = 0$

$k = mg/x = [(1.10 \text{ kg})(9.80 \text{ m/s}^2)]/(0.110 \text{ m}) = 98.0 \text{ N/m}$

b. By conservation of mechanical energy (v_0 is the maximum velocity and A is the maximum amplitude)

$U_i + K_i = Uf + Kf$

$0 + K_i = Uf + 0$

$(1/2)mv_0{}^2 = (1/2)kA^2$

$v_0 = \sqrt{kA^2/m} = \sqrt{[(98.0 \text{ N}/\text{m})(0.500 \text{ m})^2]/1.10 \text{ kg}} = 0.472 \text{ m}/\text{s}$

c. $v = v_0\sqrt{1 - x^2/A^2} = (0.472 \text{ m}/\text{s})\sqrt{1 - (2.5)^2/(5.0)^2} = 0.409 \text{ m}/\text{s}$

d. $T = 2\pi\sqrt{m/k} = 2\pi\sqrt{(1.10 \text{ kg})/(98.0 \text{ N}/\text{m})} = 0.666 \text{ s}$

$ST = 3.33 \text{ s}$

When there are no nonconservative forces, use conservation of mechanical energy.

---

16. (25 pts) You want to build a rectangular wooden box that is closed at one end that will resonate with a fundamental frequency of 262 Hz (middle C) when the temperature is 20°C.

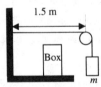

a. How long should the box be?

b. You find some wire that has a mass of 0.0200 kg and is 3.0 m long. You tie a 1.5-m section of this wire to a wall, hang a mass from the wire and set your wooden box underneath it, as shown in the figure. What mass (m) should you hang from the pulley so that the wire has the same fundamental frequency as the box?

c. Because you don't have an accurate weight, you do your best to find something with a mass that you determined in part b. However, you find that when you strike the box and pluck the string you hear a beat frequency of 4.0 Hz. If the string produces a higher pitch than the box, what is its actual frequency?

Answers:

a. At 20°C, the velocity of sound in air is given by:

$$v = 331 + 0.6(20) = 343 \text{ m/s}$$

$$f = v/4L$$
$$l - v/4f = (343 \text{ m/s})/[4(262 \text{ s}^{-1})] = 0.33 \text{ m}$$

b. $f = v/\lambda$

With $\lambda = 2L$ and $v = \sqrt{T/_{m/l)}}$, $f = \dfrac{\sqrt{T/_{(m/l)}}}{2L}$

$$f^2 = T/[4L^2(m/l)]$$
$$T = (f^2 4L^2/(m/l) = [(262 \text{ s}^{-2})4(1.5 \text{ m})^2][3.0 \text{ m})/0.00200 \text{ kg})] = 410 \text{ N}$$

T $\Sigma F = 0$

              $T - mg = 0$

              $m = T/g = (410 \text{ N})/(9.80 \text{ m/s}^2) = 42 \text{ kg}$

mg

c. $f' = f + f_{beat} = 262 + 4 = 266 \text{ Hz}$

CHAPTER 26

COLLEGE OF WILLIAM AND MARY

PHYSICS 107: PHYSICS FOR LIFE SCIENTISTS

Marc Sher, Associate Professor

THREE MIDTERMS AND A FINAL ARE GIVEN IN THIS COURSE. AN EXAMPLE OF ONE midterm and a final are presented here.

This is an introductory physics course geared to the interests of students in the life sciences. My course objectives are not just to have the students learn basic principles of physics, but to have them (1) learn to think like scientists and (2) understand the importance of physics to biology and medicine. I have a lot of biological applications in the course.

I don't expect students to remember all of the physical principles, but to understand the thought processes of physicists. I give them a Fermi question every day, and consider the ability to solve Fermi questions the most important part of the course. I'm more interested in their learning the very basic, qualitative principles than memorizing rote formulas—for example, understanding the role that Bernoulli's principle plays in transient ischemic attacks (TIAs), rather than memorizing the formula for Bernoulli's equation.

Obviously, one must present the basic formulas and do lots of problems, but by relating them to biological examples, it keeps them interested.

Important note: Calculators were not allowed on the exams—I didn't count off if the answers were left in "unreduced" terms. I also told students to use 10 m/s^2 for the gravity constant. In addition, the basic equations were given the students on the front of the exam; these equations were copied from the summary of the relevant chapters in the text.

Note that the diagrams presented are hand-drawn, exactly as they are presented in the actual examinations.

MIDTERM EXAM

1. A bowler holds a bowling ball whose mass M is 8.0 kg in the palm of his hand. The forearm and hand together have a mass m of 2.0 kg. The flexor muscle exerts a tension M; the upper arm exerts a force F on the lower arm. The relevant forces and distances are shown in the diagram (the size of the forces are not to scale, but they are all vertical). Find F and M. Compare the tension in the biceps with the weight of the ball.

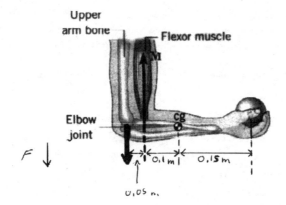

Answers:

Forces in y direction:
$$M - F - (2.0)g - (8.0)g = 0, \qquad \text{so} \qquad M = F + 100$$

Torques:
CCW: $F(.05)$
CW: $(2.0)g(.1) + (8.0)g(.25) = 22$

equating:
$F = 22/.05 = 440$ N
so $M = F + 100 = 540$ N
This is about seven times the weight of the ball, which is 80 N.

The basic method of doing all statics problems is to:

1. Draw all of the forces (including their location), including a force mg down, drawn from the center of mass of any object.

2. Add up all of the x components of the forces and set the sum to zero.

3. Add up all of the y components of the forces and set the sum to zero.

4. Choose an axis of rotation.

5. Write the torque associated with each force, and specify whether it makes the system rotate clockwise or counterclockwise.

6. Set the sum of all of the clockwise torques equal to the sum of the counterclockwise.

The resulting equations will be sufficient. All statics problems can be solved this way.

This problem has a massive forearm-hand, a massive bowling ball, and two forces on the forearm, given in the diagram. The first step is to draw the forces of gravity. There are no x components of forces, so step 2 is unnecessary. The first line is step 3. Note that $g = 10$ is used.

Next, an axis is chosen at the site of M. If one chose it at the site of F, one would get the same answer. The CCW torques are listed, as are the CW torques. They are equated, and one solves the two equations.

It is important to state clearly what one is doing: there's nothing harder for a grader than seeing $M = F + 100$ without knowing where it came from (especially if there's a simple arithmetic error—they then can't assign partial credit). Note that what is being done is clearly stated at the beginning of each line.

It is important, in doing any statics problem, to be very systematic.

2. A merry-go-round is a disk with mass M and radius R. A child with mass m is standing on the edge. The disk is spinning with angular velocity w.

 a. What is the moment of inertia of the merry-go-round-plus-child system? What is the angular momentum?

 b. The child walks to the center of the merry-go-round. What is the new angular velocity?

 c. Is the merry-go-round spinning faster, slower, or the same?

Answers:

a. $I = (1/2)MR^2 + mR^2$

 $L = I\omega$

b. The new I is $I_f = (1/2)MR^2$, because the child is on the axis. Conservation of L gives $[(1/2)MR^2 + mR^2]\Omega = (1/2)MR^2 \omega_f$, so the final angular velocity is (the Rs cancel):

 $(M + 2m)\omega/M$

c. The final angular velocity is bigger, so it spins faster.

This is a conservation of angular momentum problem—you have a spinning object, something changes, and it spins at a different rate. In conserving angular momentum, one needs the initial and final angular momenta. The primary equation for angular momentum is $L = I\omega$, so one needs the initial and final moments of inertia. The first part asks for the initial moment of inertia. Many students have trouble with this. It is important to remember that when you have two (or more) objects rotating around an axis, the total angular momentum is the sum of the angular momentum of each. The merry-go-round is a disk, and has a moment of inertia of $1/2MR^2$. The child is *not* a disk (many students make this mistake), but is a point going around a distance R from the axis, so its moment of inertia is just mR^2. When he/she walks to the center, he/she is no longer a distance R away, but a distance zero away, so there is no moment of inertia.

This problem demonstrates whether the students have a good understanding of the concept of moment of inertia. For the last part, one just has to look and see whether the final angular velocity is bigger or smaller than the initial. Because $(M + 2m)/M$ is always bigger than one, it is bigger, and the merry-go-round spins faster.

3. A 2.0-kg mass is hung from the end of a vertical spring. The spring is stretched 10 cm from its equilibrium position. The mass of the spring is negligible.

a. What is the spring constant, k?

b. The spring is now put on a frictionless, horizontal table, and the same mass is attached. It is pulled back 20 cm and released. What is the period of the oscillation?

c. How many cycles does it make in 10 s?

d. What is the total energy of the system?

e. When $x = 5$ cm, what is the velocity of the mass? (*Hint:* first, find the potential energy of that point, then use the answer to part d.)

Answers:

a. Force up is kx, force down is mg, so:

$$kx = mg$$

so:

$$k = mg/x = (2)(10)/(.1) = 200 \text{ N/m}$$

b. $T = 2\pi(m/k)^{1/2} = (6.3)(2/200)^{1/2} = .63$ s

c. $10/6.3 = 16$ cycles (approx.)

d. $E = 1/2 \, k \, A^2 = (1/2)(200)(.2)^2 = 4$ J

e. $PE = 1/2 \, k \, x^2 = (1/2)(200)(.05)^2 = .25$ J

so: $KE = E - PE = 3.75$ J

$KE = 1/2 \, m \, v^2$

so:

$$3.75 = (1/2)(2)v^2$$

or

$$v = 1.9 \text{ m/s}$$

This is a straightforward spring problem, but there are several possible pitfalls. With any vertical spring, there will be gravity competing against the restoring force. Because they balance, one must have $kx = mg$, which solves the problem. The potential pitfall here is units. The amount of stretching is given in centimeters, and yet the mass is given in kilograms. So, you must convert everything (at the beginning of the problem!) to meters.

For part b, the spring is now horizontal, so gravity is irrelevant, and the spring oscillates back and forth. The answer is a plug-in to the formula for period.

Part c can be tricky; students can plug in to a formula, but this part makes sure you know what a "period" of oscillation is. If there is a period of .63 s, then in 10 s it oscillates 10/.63 times. (If you

don't see that, then think about if it had a period of 1/3 s—then in 10 s it would oscillate 30 times.)
Part d is a plug-in to the formula for total energy of an oscillator—again, units can cause problems.
Finally, for part e, it is necessary to understand that an oscillation is an interplay between kinetic
and potential energy.

4. An air hose is blowing air through the Venturi tube as shown. The radius of the tube
at points A and C is 2 cm, and at point B is 1 cm. The tube is connected in three
places to tubes containing mercury. The density of mercury is 13,600 kg/m³, that of
air is about 1 kg/m³.

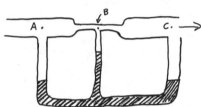

a. The velocity of the air coming out of the air hose is 30 m/s. What is the velocity
of the air at point B?
b. What is the pressure difference between points A and B?
c. What is the height difference between the first and second mercury columns?

Answers:

a. $A_C v_C = A_B v_B$

so

$v_B = (A_C/A_B)v_C = (r_C/r_B)^2 v_C$

so

$v_B = (2/1)^2 (30) = 120$ m/s

b. From Bernoulli:

$p_A + (1/2)\rho v_A^2 = p_B + (1/2)\rho v_B^2$

so

$p_A - p_B = (1/2)\rho (v_B^2 - v_A^2) = (1/2)(1)(120^2 - 30^2) = 6750$ Pa

c. From Bernoulli: the pressure difference is ρgh, where ρ is the density of mercury.

So:

$6750 = (13,600)(10)h$

or h is about .05 m

This is a very good problem. It covers the subjects of flow rates and Bernoulli's equations very thoroughly. Students should know that fluid (or gas) will flow more quickly through a narrower pipe,
and the formula is that the area times the velocity is constant. Because the pipe is cylindrical, the
cross-sectional area is pr^2. Here is a case in which it is better to do the problem algebraically first.
There is no need to get the actual area of the pipe, only the ratio of areas, and thus the ps drop out
(makes the algebra easy).

Part b asks for the pressure difference between points A and B. Because they are at the same height, there is no difference in ρgh, so the only effect must be from the velocity. Note that the density here is that of air, not mercury, because air is flowing through the tube. This is a very straight-forward application of Bernoulli's principle: lower pressure means greater velocity.

For part c, you are asked about the height of the mercury, so the relevant density is that of mercury. Because the mercury isn't moving, the only effect on the pressure is due to ρgh.

In this problem, it is more important to have a basic understanding of the physical concepts than to just memorize formulas (or read them off the front sheet).

FINAL EXAM

Do the homework!!! Do several problems extra for every one assigned. If you don't understand it, ask the instructor for help. Too few students ask for help, get far behind, and run into trouble.

Part A: Short-answer questions

(Answer in the space provided, if possible. Use back if necessary.)

A1. Two charges, $+Q$ and $-Q$, are shown. At each of the five points, A, B, C, D, and E, draw an arrow showing the direction of the electric field at that point.

Answer:

A2. Rank the following in order of *increasing* wavelength: radio, ultraviolet, blue, gamma ray, red, infrared.

Answer:
gamma ray
ultraviolet
blue
red
infrared
radio

A3. If you look at the spectral lines of hydrogen coming from the left side of the sun, they are shifted slightly to the red. If you look at those coming from the right side, they are shifted slightly to the blue. What can you conclude from this?
Answer: The sun is rotating, so one side is moving away from us, the other is moving towards us, and the Doppler shift changes the frequencies.

A4. In class, the professor sent light through two crossed polarizers, showing that no light emerged. He then *inserted* a third polarizer between the two, and light reappeared. Explain how this could happen.
Answer: When light goes through the first polarizer, it emerges polarized in a certain direction. If the second polarizer is perpendicular, no light gets through. But if the second is at an angle of, say, 45°, then some light does get through, and it emerges polarized in the direction of the second. If the third polarizer is at 45° to the second, then light will get through (even though the third and first are perpendicular).

A5. Thorium, 232Th, undergoes alpha decay. Identify the daughter nucleus (by "identify," I mean give the charge, Z, and the nucleon number, A). The daughter nucleus then undergoes beta-minus decay. Identify the granddaughter nucleus.
Answer: Alpha decay lowers Z by 2 and A by 4, so the daughter has $Z = 88$, $A = 228$. Beta-minus decay increases Z by 1 and doesn't change A, so $Z = 89$, $A = 228$. The element is Ac.

A6. In terms of the curve showing binding energy/nucleon versus A, explain why fusion weapons are much more powerful than fission weapons.
Answer: The difference between the hydrogen and helium BE/nucleon is much greater than the difference between uranium and its fission products.

Most students drew the curve, although this isn't required as part of the answer.

A7. Explain the difference between a real image and a virtual image.

Answer: In a virtual image, the rays of light do not actually emanate from the image, whereas they do in a real image.

A8. An elderly bank president needs to hold the financial pages of the paper at arm's length, 50.0 cm, to read them clearly. What should the focal length of an eyeglass lens be to allow her to read at the more comfortable distance of 25 cm? (Ignore the distance from the glasses to the cornea.)

Answer: The near point is at 50 cm, so you want an image at 60 cm, so $d_i = -60$. The object distance is 25 cm. So:

$1/f = 1/25 + 1/(-50) = 1/50$

so:

$f = 50$ cm

A9. You have a coil connected to an ammeter and move a magnet into the coil. Between $t = 0$ and $t = 1$ s, you move the magnet into the coil, between $t = 1$ and $t = 2$, you let it sit there, and between $t = 2$ and $t = 3$, you slowly remove it. During these 3 s, when is current flowing in the coil?

Answer: Current flows when the magnetic field changes, so current flows between 0 and 1, and between 2 and 3.

Part B

B1. Five resistors are connected to a 20-V battery as shown.

a. Find the equivalent resistance and the current coming out of the battery.

b. Find the current and total power going through *each* of the five resistors. If they were lightbulbs, which would be brightest?

Answers:

a. The 2- and 10-Ω resistors combine to make a 12-Ω; the two 3-Ω resistors make a 6-Ω. A 6-Ω resistor in parallel with a 12-Ω has resistance given by $1/(1/12) + (1/6)$, or 4 Ω. The two 4-Ω resistors then have an 8-Ω equivalent resistance. The current coming out of the battery is $V/R = 20/8 = 2.5$ A.

b. All of the current goes through the 4-Ω resistor, so $I = 2.5$ A and $P = I^2R = (2.5)^2(4) = 25$ W. The voltage drop across the 4-Ω is $IR = (2.5)(4) = 10$ V, so the voltage drop left, across the 3 + 3 or the 2 + 10 resistors is 10 V. The current through the 3 + 3 resistors is then $(10)/(3 + 3) = 1.667$ A. The current through the 2 + 10 resistors is $(10)/(2 + 10) = .833$ A. These add up to 2.5, as they should. To get the power, use $P = I^2R$.

3 Ω: $(1.667)^2(3) = 8.33$ W
2 Ω: $(.833)^2(2) = 1.4$ W
10 Ω: $(.833)^2(10) = 7$ W

This is a very standard "resistor" problem, and yet it causes students quite a bit of difficulty. Students tend to try to just "find the right formula," without thinking about what is going on; that sometimes works, but not here. In any problem with many resistors, the first thing to do is to use the formulas for resistors in series and parallel to simplify the problem down, hopefully, to a single resistor. The 2- and 10-Ω resistors are in series, so they are equivalent to a 12-Ω. The two 3-Ω resistors are also in series, so they are equivalent to a 6-Ω. Now, you have a new problem, with only three resistors. The 6-Ω and 12-Ω are in parallel—the formula says that they are equivalent to a 4-Ω. Finally, you now have two 4-Ω resistors in series, which is equivalent to a single 8-Ω resistor. For a single 8-Ω resistor and a 20-V battery, the current is $I = V/R$, or 2.5 A. Note that you reduce it all down *first*, then use the formula.

This does *not* mean that the current through all of the resistors is 2.5 A. Look at the picture. If 2.5 A leaves the battery, it will go through the 4-Ω resistor, and then it will divide up. So you *won't* have 2.5 A going through the others.

One way to proceed is to note that the 12-Ω equivalent resistor has twice the resistance as the 6-Ω, so (because the voltage is the same—they're in parallel) the current through the 12-Ω equivalent resistor will be one-half of the 6-Ω. Thus, the 2.5 A divide up, with one-third going through the 2- and 10-Ω resistors and two-thirds through the 3-Ω resistors.

Another way, given in the answer, is to note that the voltage drop across the 4-Ω resistor is *IR*, or $(2.5)(4) = 10$ V, so the 20 V that you start with drop to 10 V after going through the 4-Ω. That means that the voltage drop across the 12- and 6-Ω equivalent resistors is 10 V. Dividing by *R* gives the current, 10/12 for the 12-Ω and 10/6 for the 6-Ω. Finally, that gives you the current going through each resistor, and I^2R gives the power. The 4-Ω will be the brightest.

Go over this problem carefully. It really is a good example of series and parallel circuits, and if you understand it, you should be able to do most circuit problems.

B2. A 12-V battery is connected to a parallel plate capacitor with a capacitance of 1.0×10^{-6} F. The plates of the capacitor are separated by 6 mm.

a. Determine the area of the plates.

b. Determine the charge on each plate

c. Determine the magnitude of the electric field between the plates.

d. An electron is released from the negative plate. How much kinetic energy does it have when it reaches the positive plate?

e. The battery is disconnected and a sheet of paraffin (dielectric constant 2.2) is slipped between the plates. How do your answers to parts b and c change? What is the voltage across the plates?

Answers:

a. $C = A\varepsilon_0/d$

so

$A = Cd/\varepsilon_0 =$

$(1 \times 10^{-6})(.006)/(8.8 \times 10^{-12}) = 650 \text{ cm}^2$

b. $Q = CV = 1.2 \times 10^{-5} \text{ C}$

c. $E = V/d = (1.2 \times 10^{-5})/(6 \times 10^{-3}) = .002 \text{ N/C}$

d. $KE = qV = (1.6 \times 10^{-19})(12) = 1.92 \times 10^{-18} \text{ J}$

e. Q is unchanged, C increases by 2.2, so V decreases by 2.2, so E decreases by 2.2.

> The first three parts of this problem are basic plug-ins. There is a formula for the capacitance in terms of the area and separation of parallel plates, the charge on a capacitor, and the electric field in a capacitor. Be careful with units—the separation is 6 mm.
>
> The last two parts are more challenging. When an electron crosses a potential difference, it picks up kinetic energy. The potential energy difference is qV, and that turns into kinetic energy as the electron moves. The q is the charge on the electron (*not* the charge on the plates).
>
> Here is a problem in which the understanding of the meaning of "voltage" is important. In the last part, the battery is disconnected. If that happens, then the charge on the parallel plates can't go anywhere, so Q is fixed. Looking at the formula for capacitance, you see that C increases by 2.2, and because $Q = CV$, and Q is fixed, then V must decrease by 2.2. Suppose the battery had *not* been disconnected. Then the voltage would always be 12 V, and because C increases by 2.2, Q would increase by 2.2.

B3. An electron is accelerated through a potential difference of 100 V into a magnetic field of 1.0 T. The magnetic field is perpendicular to the velocity of the electron. The mass of an electron is 9.1×10^{-31} kg, and its charge is 1.6×10^{-19} C.

a. Determine the velocity of the electron as it enters the field.

b. Explain why the electron, when it enters the magnetic field, travels in a circle. Determine the radius of the circle.

c. What is the period of its motion?

d. If the magnetic field points into the paper, would the electron travel clockwise or counterclockwise as viewed from above?

Answers:

a. $KE = qV = 1.6 \times 10^{-17} \text{ J} = 1/2 mv^2$

So

$v^2 = 2 KE/m = .36 \times 10^{14}$

so

$v = 6 \times 10^6 \text{ m/s}$

b. The force is perpendicular to the velocity, so if the acceleration is perpendicular to the velocity, it goes in a circle.

$F = qvB = ma = m(v^2/r)$

so

$r = mv/qB = 3.4 \times 10^{-5}$ m

c. $P = 2\pi r/v = (6.28)(3.4 \times 10^{-5})/(6 \times 10^6) = 3.5 \times 10^{-11}$ s

d. Clockwise.

> This problem requires a good understanding of the meaning of volts and motion in a magnetic field. When an electron is accelerated through a potential energy of 100 V, it gets a kinetic energy of 100 electronvolts (eV), or (either using a conversion factor or just the formula $KE = qV$) 1.6×10^{-17} J. Once you have the kinetic energy, getting the velocity is simple. When a particle moves in a magnetic field, the force is always perpendicular to the velocity, and thus charged particles move in circles. Using the formula for the force on a charge in a magnetic field ($F = qvB$) and for a charge going in a circle ($F = mv^2/r$), one can get the radius.
>
> Note that this is essential in understanding mass spectroscopy. If you know a particle's speed and its radius, then you know the period. The direction is determined by the right-hand rule. Remember, though, that the particle is an electron, which has negative charge. So, when you do the right-hand rule, you then reverse the result for the electron.

B4. An object 5 cm high is placed a distance D in front of a concave mirror with focal length 20 cm. For each of the following values of D, find the location and height of the image, and determine its characteristics (real, virtual, erect, inverted, bigger, smaller). For cases b, c, and e, draw a ray diagram.

   a. $D = 60$ cm

   b. $D = 10$ cm

   c. $D = 30$ cm

   d. $D = 20$ cm

   e. $D = 20$ cm, but now assume the mirror is convex with focal length 20 cm

Answers:

$(1/d_i) = 1/f - 1/d_o = 1/20 - 1/d_o$; $m = -d_i/d_o$

a. $1/20 - 1/60 = 1/30$

   so

   $d_i = 30$ cm

   $m = -30/60 = -1/2$

   (inverted, smaller, height 2.5 cm)

b. $1/20 - 1/10 = -1/20$

   so

   $d_i = -20$ cm

   $m = 20/10 = 2$

   (erect, larger, height 10 cm)

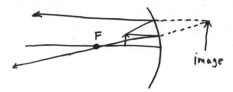

c. $1/20 - 1/30 = 1/60$

   so

   $d_i = 60$ cm

   $m = -60/30 = -2$

   (inverted, larger, height 10 cm)

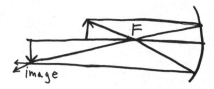

d. d_i is infinite, no image.

e. f is negative.

   $1/d_i = -1/20 - (1/20) = -1/10$

   so

   $d_i = -10$ cm

   $m = 1/2$

   (erect, smaller, height 2.5 cm)

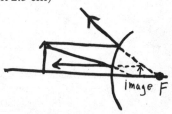

An important part of optics is the ability to locate an image, given an object and a lens or mirror. Students should look at all of the standard situations. There are only two main formulas:

$1/f = 1/d_i + 1/d_o$ and $m = -d_i/d_o$

but the applications are very wide. Students should be able to look at all possible values for f and d_o and be able to trace rays, locate the image, and so on.

   This problem gives essentially all of the main cases involved for an object and a single mirror. The first four parts have a concave mirror, and the object is either very far away, closer but outside the focal point, inside the focal point, and at the focal point. Note that if an object is at the focal point, the rays don't focus, and there is no image. In studying this material, you should be able to do the same thing for a single lens, concave or convex.

B5a. A very fine wire is placed between two flat glass ($n = 1.5$) plates as shown. Light with $\lambda = 600$ nm falls (and is viewed) perpendicular to the plates, and a series of bright and dark bands is seen. A total of 25 bright lines is seen. What is the thickness of the wire?

B5b. A soap bubble appears green ($\lambda = 540$ nm) at the point on its front surface nearest the viewer. What is its minimum thickness? ($n = 1.35$ for soap)

Answers:

B5a. $t = (m + 1/2)\lambda_{film}/2$. The film is air, so the wavelength is 600 nm. This gives $t = 12.75\ (600\text{ nm}) = 7.5 \times 10^{-6}$ m.

B5b. Same formula, now the film is soap, so the wavelength is $540/1.35$ nm = 400 nm. The minimum thickness is $m = 0$, so $t = 400/4 = 100$ nm.

This is a thin-film problem, but there are some possible pitfalls. When the wire is put between two glass plates, the "film" is just a film of air, between the plates. Although you can memorize formulas to solve the problem, it's easier to just think about it. The light that reflects off the bottom plate travels an extra distance compared with light that reflects off the upper plate. If D is the gap at the point the light reflects, then the extra distance is $2D$. If this matches an integral number of wavelengths, there will be destructive interference (destructive because of the extra 1/2 wavelength phase shift on reflection). So you see a dark line if $2D = m\lambda$, where m is any integer. You see a bright line if $2D = (m + 1/2)\lambda$. So, if you see 25 bright lines, the thickness of the wire is given by D, with $m = 25$. Because the film is *air*, the relevant wavelength is the wavelength in air. There are lots of formulas involving thin films (some have the 1/2, some don't, some are $2D$, some are D), and it's often safer and easier just to think about what interference means, as in the preceding. (If you forget the 1/2, your answer will be pretty close anyway.)

For part b, there is another thin film, but this time it is soap, with $n = 1.35$, so the relevant wavelength is smaller by that factor. There are many possible thicknesses that will give green (corresponding to a "bright line"), each for a different m, but the minimum thickness is $m = 0$.

B6. A fission reactor is hit by a nuclear weapon, causing 5.12×10^6 curies (Ci) of Strontium-90 to evaporate into the air. The Strontium-90 falls out over an area of 10,000 km². The half-life of Strontium-90 is 28.7 years.

a. How long will it take the activity of the Strontium-90 to reach the agriculturally "safe" level of 2.0 μCi/m²?

b. At this "safe" level, how many atoms of Strontium-90 decay every second in every square centimeter? [Note: 1 Ci = 3.7×10^{10} becquerels (Bq).]

c. Strontium-90 is an alpha emitter. Once it has all fallen out of the atmosphere (and is in the soil), why would it be safe to stand in the area (with shoes on), but dangerous to eat food grown in the soil?

Answers:

a. 10,000 km^2 is 10^{10} m^2, so the activity is 5.12×10^{-4} Ci/m^2, or 512 μCi/m^2. To drop to the safe level, a reduction of 256 is needed, or eight half-lives. This takes 28.7 (8) = 230 years.

b. 2×10^{-6} Ci = 7.4×10^4 Bq. This is 74,000 disintegrations per second per square meter, or 7.4/cm^2.

c. Alphas are short range and are thus only dangerous if taken internally.

An understanding of half-life is crucial in biology and medicine. A difficulty many students had with this problem was with the units. The Strontium-90 falls over 10,000 square *kilometers*, and yet the question asks about square meters. Because 10,000 km^2 is 10^{10} m^2, the 5.12×10^6 Ci fall over 10^{10} m^2, or 5.12×10^{-4} Ci/m^2. Then you have to convert to microcuries, giving 512 μCi/m^2. From that point, it's easier, but if you messed up with the units, the problem becomes much more difficult. To go from 512 to 2 requires eight steps (512, 256, 128, 64, 32, 16, 8, 4, 2), or eight half-lives—230 years. Note that if you mess up the units, the numbers won't divide evenly—that should always be a clue to go back and check more carefully.

For part b, it is essential to know what a curie is. Much of understanding radioactivity involves remembering the units—a curie is 3.7×10^{10} Bq, or disintegrations, per second. Finally, the last part tests the student's understanding of the biological differences between alpha and beta radiation.

PART FOUR

FOR YOUR REFERENCE

CHAPTER 27

A GLOSSARY OF PHYSICS

Absorption The transfer of energy from an electromagnetic wave to matter.

Acceleration Vector describing the rate of change of velocity.

Alpha ray Helium nucleus possessing high kinetic energy; often produced by a nuclear reaction.

Amplitude Size of the disturbance created by a passing wave or the size of a field.

Angular acceleration Vector describing the rate of change of an angular velocity.

Angular momentum Vector that is the product of angular velocity times the moment of inertia.

Angular speed Scalar describing the rate of rotation.

Angular velocity Vector describing an object's rate of rotation and the orientation of its rotation axis.

Atom Smallest unit into which an element can be divided without losing its identity as that element.

Atomic mass A number approximately equal to the number of neutrons and protons in an atom's nucleus.

Atomic number Number of protons in the nucleus of an atom; also the electrical charge of the nucleus in units of the magnitude of the electron's charge.

Beta ray Electron with a large kinetic energy, often emitted in a nuclear reaction.

Big bang The primordial explosion of a highly compact universe, which caused the expansion of the universe.

Black body An idealized object that absorbs all electromagnetic radiation falling on it and emits all the energy it absorbs.

Black-body curve Curve of the distribution of power emitted by a radiating black body.

Black hole The end result of the implosion of an extremely massive star. It is an object from which no light, no information, no radiation of any kind, escapes.

Bohr atom First modern model of the atom, which predicted discrete atomic spectra.

Carnot efficiency Theoretical limit to the efficiency of an ideal heat (Carnot) engine.

Center of mass The point in or near a rigid object that moves as if all the object's mass were concentrated at it.

Charge Quantity of electricity in an object.

Closed universe A universe finite and without boundaries. A universe with density above the critical value is necessarily closed.

Complementarity Principle that wave-particle duality is central to quantum mechanics.

Conduction Flow of heat or electric current through matter.

Conductor Substance capable of conducting the flow of heat or electric current.

Convection Movement of hot masses in a gas or liquid causing heat transfer.

Cosmology Study of the origin, present state, and possible futures of the universe.

Current Rate of flow of a quantity, especially electrical charge or heat.

de Broglie wavelength Wavelength of the quantum mechanical wave associated with any object that has momentum.

Diffraction Spreading and interference that occur after a wave passes an obstacle.

Doppler shift Frequency change between a wave source and an observer with nonzero relative velocity.

Duality Principle that quantum objects sometimes behave as waves and sometimes as particles.

Electromagnet A coil that produces a magnetic field when current flows through it.

Electromagnetic induction Effect produced by a changing magnetic field.

Electromagnetic radiation Energy in the form of rapidly fluctuating electric and magnetic fields and including visible light in addition to radio, infrared, ultraviolet, x-ray, and gamma ray radiation.

Electromotive force (emf) Non-Coulomb force, such as an inductive force, on a charged particle.

Electron Small-mass subatomic particle with negative charge.

Emission The radiation of light from an excited atom.

Energy Conserved quantity describing the capacity of a system to do work.

Entropy Measure of disorder in a thermodynamic system.

Escape velocity The velocity necessary for an object to escape the gravitational pull of another object.

Event horizon An imaginary boundary surrounding a collapsing star or black hole. Within the event horizon, no event can be communicated to the outside.

Field Quantity defined at every point in a system.

Fluid Matter that flows, assuming the shape of its container but not holding that shape. May be a liquid or a gas.

Flux Quantity of matter that crosses a unit area in a unit time.

Force Vector that describes the direction and magnitude of a push.

Free fall Motion of an object subject to gravity alone.

Frequency The number of wave crests that pass a given point per unit of time. By convention, this is measured in *hertz* (equivalent to one crest-to-crest cycle per second and named in honor of the nineteenth-century German physicist Heinrich Rudolf Hertz).

Friction Force acting on a moving object in contact with another object. Friction transforms mechanical energy into heat.

Gamma ray Electromagnetic radiation emitted at frequencies higher than those of X rays and light; often emitted in nuclear reactions.

Half-life The time required for half of a quantity of identical unstable nuclei to decay.

Heat Energy property of matter that flows from relatively hot to relatively cold objects.

Heat engine Device that converts a fraction of the heat flowing into it to mechanical work or converts a fraction of mechanical work to heat.

Impulse Product of a force multiplied by its duration and equal to the change in momentum produced by the force.

Inertial frame Frame of reference from which objects not acted on by forces are seen to move with constant velocities.

Insulator Matter that does not conduct currents.

Interference Adding (constructive interference) or canceling (destructive interference) of amplitudes that occurs when two or more waves come together.

Isotope Atomic nuclei with the same number of protons but different numbers of neutrons are different isotopes of a given element.

Kepler orbit Elliptical trajectory of a planet around a star, which is at one of the foci.

Light-year The distance light travels in one year: approximately 5.88 trillion mi (9.46 trillion km). In the vastness of space, astronomers use the light-year as a basic unit of distance.

Longitudinal wave Polarization in which wave motion is parallel to the wave's direction.

Mass The measure of inertia; in effect, the amount of matter in an object.

Mechanics Branch of physics describing the motions of objects and the forces acting on these motions.

Molecule Unit of matter made up of one or many atoms bound together.

Moment of inertia Measure of the rotational inertia of an object.

Momentum Vector product of mass times the velocity of an object.

Natural frequency Frequency of an oscillator free from any external interference.

Neutron Massive, uncharged particle found in atomic nuclei.

Nuclear binding energy Energy required to separate the neutrons and protons in a nucleus from each other.

Nuclear fission A nuclear reaction in which an atomic unit splits into fragments, thereby releasing energy. In a fission reactor, the split-off fragments collide with other nuclei, causing them to fragment, until a *chain reaction* is under way.

Nuclear force The force holding the neutrons and protons in a nucleus together.

Nuclear fusion A nuclear reaction that produces energy by joining atomic nuclei. Although the mass of a nucleus produced by joining two nuclei is less than that of the sum of the original two nuclei, the mass is not lost; rather, it is converted into energy.

Nucleon Collective term for a neutron or proton.

Nucleus Positively charged center of an atom containing most of its mass.

Orbit The closed trajectory of a planet or satellite.

Pendulum A mass (the *bob*) supported on a rod or string and free to swing (oscillate).

Period Duration of one oscillation cycle or one orbit.

Photon One quantum of light.

Planck's constant The quantity (of length, time, energy) below which quantum effects are important.

Polarization Direction of wave motion relative to the direction of propagation.

Potential Electrical quantity describing the electrical potential energy per unit charge that a charge would have if placed at the point described.

Potential difference Difference between the electrical potentials at two different points.

Power Rate of energy transfer.

Pressure Force per unit area a fluid exerts on its container or on an object immersed in it.

Proton Massive, charged particle in atomic nuclei. The hydrogen nucleus is a proton.

Radiation Propagating waves, usually electromagnetic.

Radioactivity Emission of high-energy photons or charged particles by unstable atomic nuclei.

Redshift An increase in the wavelength of electromagnetic radiation emitted by a celestial object as the distance between it and the observer increases. The name derives from the fact that lengthening the wavelength of visible light tends to redden the light that is observed.

Refraction Change of direction of light rays crossing from one transparent medium into another.

Resistor Electric circuit component the operation of which is described by Ohm's law.

Resonance The large-amplitude response of an oscillator to a force periodically applied at or near the natural frequency of the oscillator.

Rotational inertia Property of resisting change in the rate of rotation of an object.

Scalar Quantity with magnitude and units, but without direction.

Semiconductor Material possessing characteristics of a conductor and an insulator, through which electrons can move freely, but in relatively low quantity.

Sine wave Mathematical function that describes the position of a pendulum as a function of time or other periodic motions, especially simple waves.

Solenoid An electromagnet wound around an air core.

Space-time The four-dimensional frame of reference in which special and general relativity are considered.

Spectrum A graphic plot (as of electromagnetic radiation) of intensity as a function of wavelength, frequency, or energy.

Standing wave The superposition of two or more propagating waves producing a pattern that oscillates in amplitude but does not otherwise move.

Superconductor An electrical conductor offering zero resistance at sufficiently low temperature.

Superposition Simultaneous production of two or more waves or motions to produce a new wave or motion.

Temperature Measure of the tendency for heat to flow from one object to another.

Tension Forces tending to pull the molecules of a solid closer together.

Thermal expansion Increase in size that occurs in most condensed matter with increase in temperature.

Thermodynamics Branch of physics that studies heat flow, temperature changes, and phenomena associated with heat flow and temperature change.

Thought experiment A systematic hypothetical or imaginary simulation of reality; used as an alternative to actual experimentation when such experimentation is impractical or impossible.

Time dilation The lengthening of time (as perceived by an external observer) of an object at relativistic speed.

Torque Product of a force multiplied by the perpendicular distance from its line of action to a rotation axis.

Trajectory Path followed by a moving object.

Transverse wave Polarization of a wave that is perpendicular to its direction of propagation.

Uncertainty principle The impossibility of simultaneously determining the exact values of all physical variables at the quantum level (that is, at scales smaller than Planck's constant).

Vector Quantity with direction and magnitude.

Voltage Electric potential.

Wave Disturbance in a medium that propagates through the medium and transmits energy.

Wavelength The distance between two adjacent wave crests (high points) or troughs (low points). By convention, this distance is measured in meters or decimal fractions thereof.

Weight Gravitational force acting on an object near a star or planet. Not to be confused with mass.

Work Product of force multiplied by the distance moved parallel to the force. Work is equal to the energy transferred to the object moved.

X ray Short-wavelength, high-energy electromagnetic radiation.

CHAPTER 28

RECOMMENDED READING

Adams, Steve. *Relativity: An Introduction to Space-Time Physics.* Taylor & Francis, 1997.

Biehle, Garrett. *MCAT Physics Book.* Nova Press, 1997.

Cottingham, W. M. *An Introduction to Nuclear Physics.* Cambridge University Press, 1986.

Davies, P. C. W. *About Time: Einstein's Unfinished Revolution.* Touchstone, 1996.

French, A. P., and Edwin F. Taylor. *Introduction to Quantum Physics.* W. W. Norton, 1978.

Halpern, Alvin. *3000 Solved Problems in Physics (Schaum's Solved Problems Series).* McGraw-Hill, 1988.

Hecht, Eugene, and Frederick J. Bueche. *Schaum's Outline of Theory and Problems of College Physics,* 9th ed. McGraw-Hill, 1997.

Islam, M. Asad, et al. *Physics I (Test Yourself).* NTC Publishing Group, 1998.

Mansfield, Michael, and Colm O'Sullivan. *Understanding Physics.* John Wiley & Sons, 1998.

McDermott, Lillian C., et al. *Physics by Inquiry: An Introduction to Physics and the Physical Sciences,* 2 vols. John Wiley & Sons, 1996.

Miller, Arthur I. *Albert Einstein's Special Theory of Relativity: Emergence (1905) and Early Interpretation (1905-1911)*. Springer Verlag, 1997.

Mills, Robert. *Space, Time, and Quanta: An Introduction to Contemporary Physics*. W. H. Freeman, 1994.

Oman, Robert, and Daniel Oman. *How to Solve Physics Problems*. McGraw-Hill, 1997.

Purrington, Robert D. *Physics in the Nineteenth Century*. Rutgers University Press, 1997.

Seaborn, James B. *Understanding the Universe: An Introduction to Physics and Astrophysics*. Springer Verlag, 1997.

Taylor, Edwin F., et al. *Spacetime Physics: An Introduction to Special Relativity*. W. H. Freeman, 1992.

Note: This index covers Part One: Preparing Yourself and Part Two: Study Guide, pages vii-93 of the text. Material in the sample exams, pages 96–303, is not indexed.